Best wishes!
[illegible] & Burton

DRAWN TO THE LAND

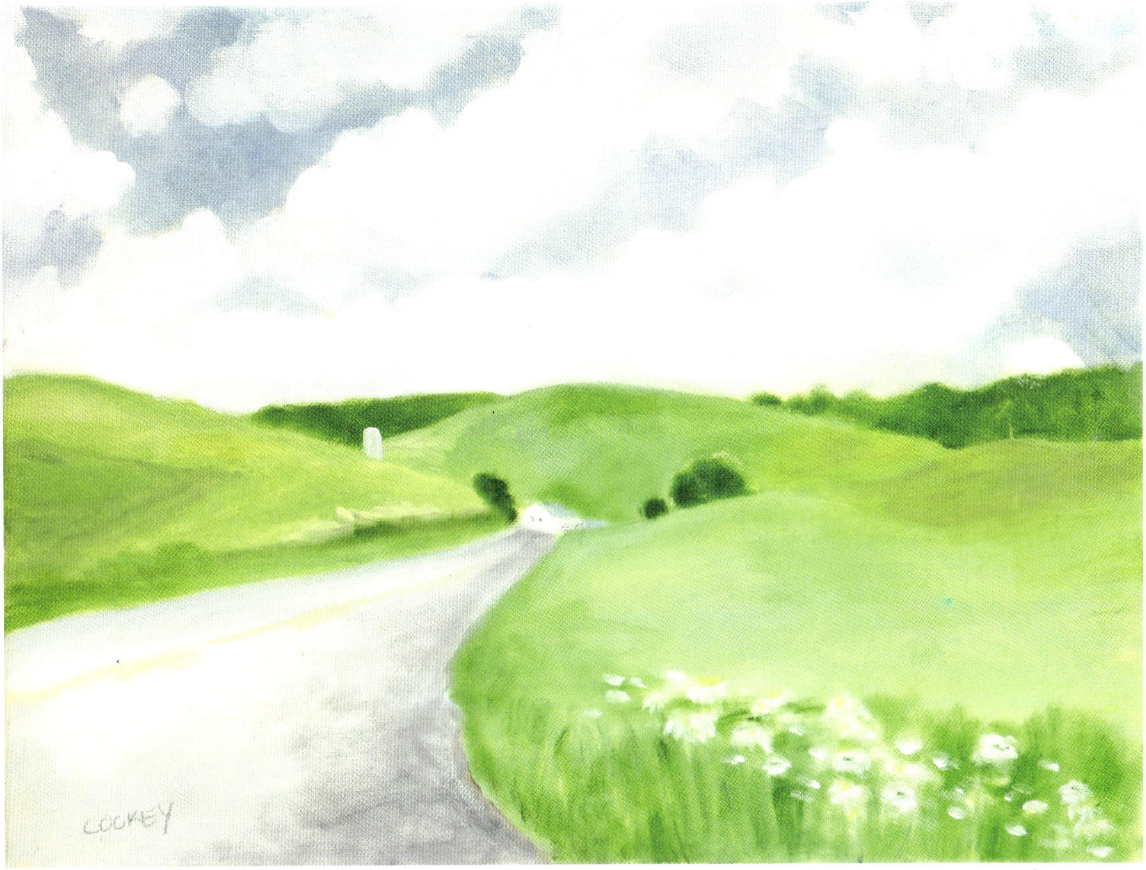

The Romance of Farming

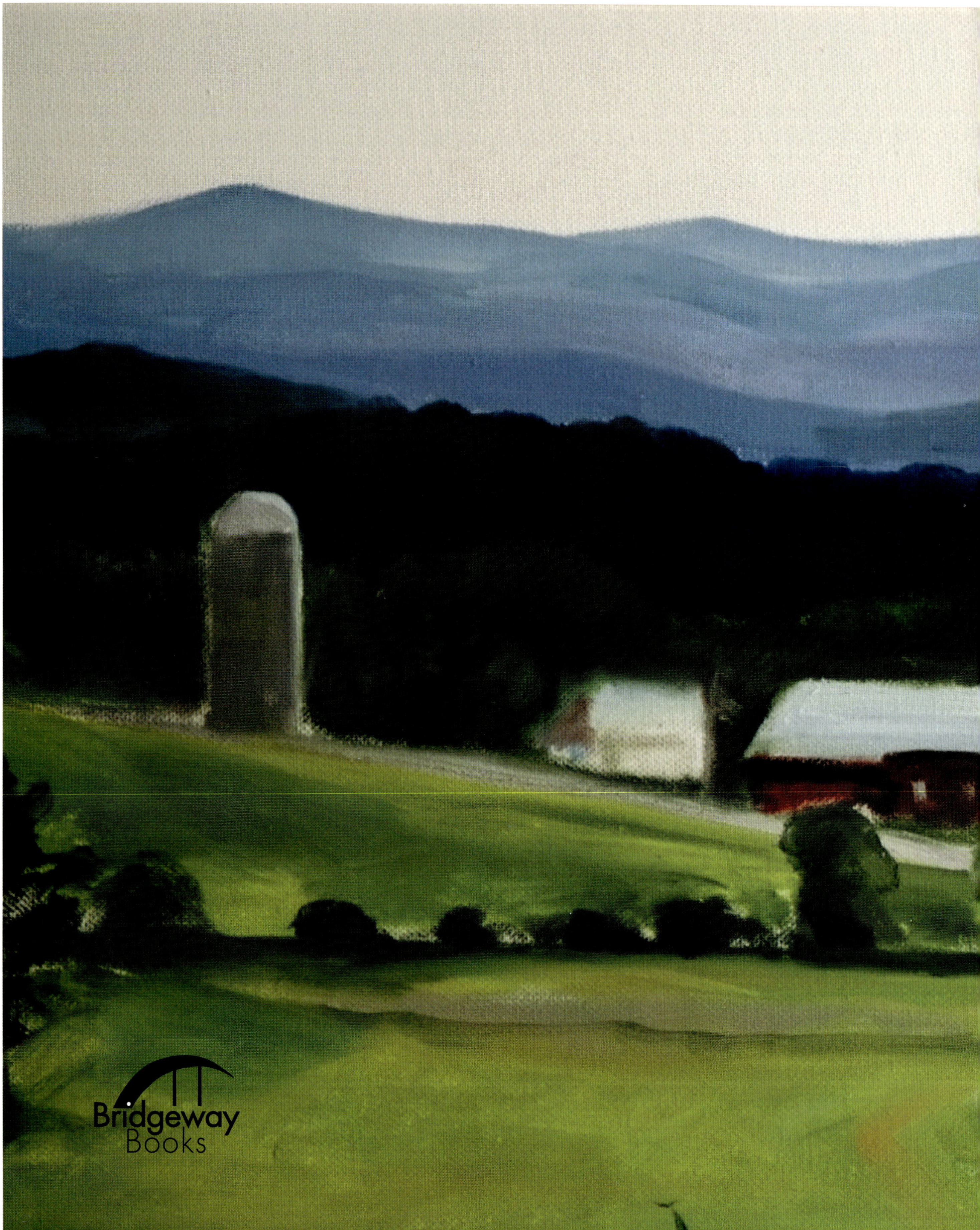
Bridgeway
Books

DRAWN TO THE LAND

The Romance of Farming

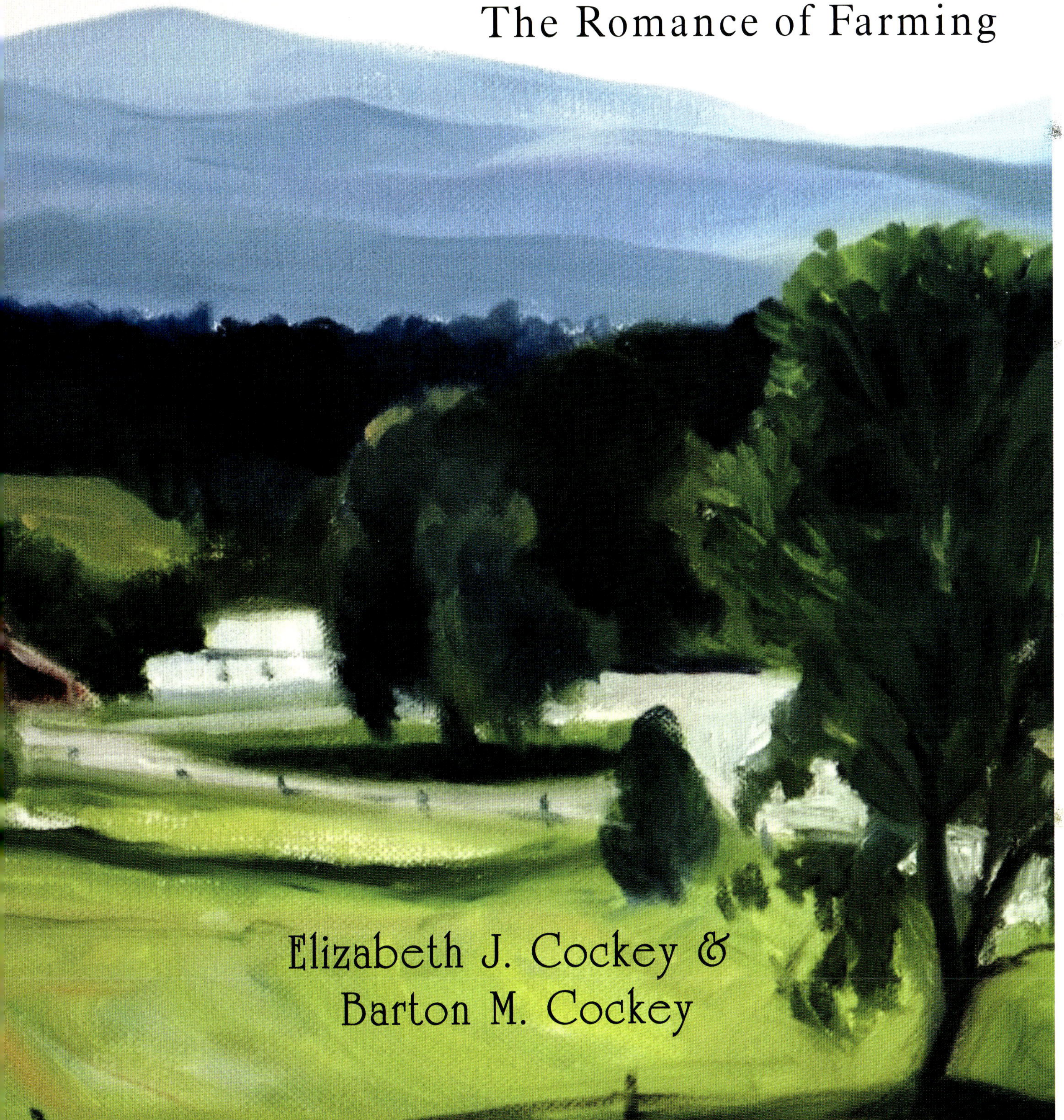

Elizabeth J. Cockey &
Barton M. Cockey

Drawn to the Land: The Romance of Farming
Published by Bridgeway Books
P.O. Box 30071
Austin, Texas 78755

For more information about our books, please write us, e-mail us at info@bridgewaybooks.net, or visit our web site at www.bridgewaybooks.net.

Library of Congress Control Number: 2009942876

ISBN-13: 978-1-934454-40-4
ISBN-10: 1-934454-40-0

Photography by F. Paul Galeone Photography

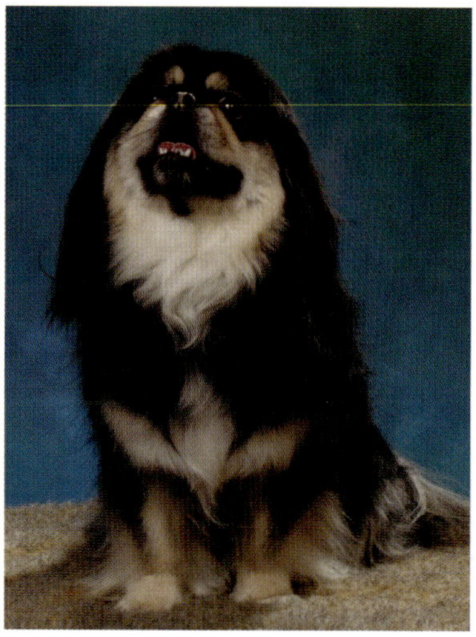

In memory of Piggy Pie, beloved companion

11/05/03–09/08/09

10 9 8 7 6 5 4 3 2 1

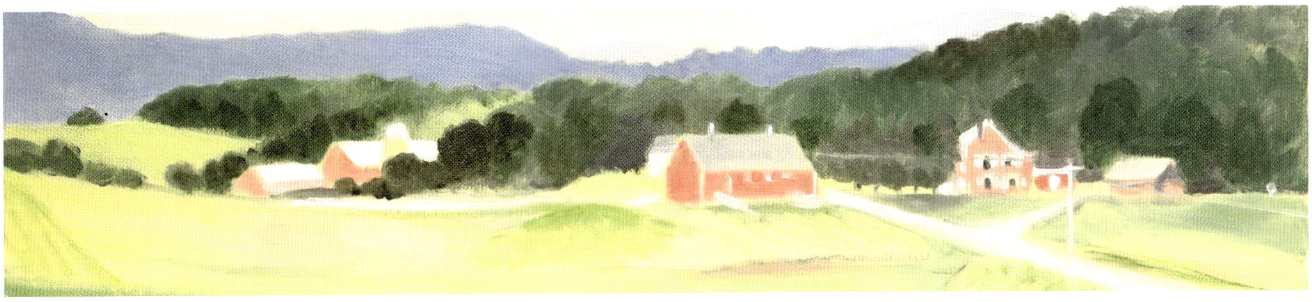

CONTENTS

INTRODUCTION

To God and home and country
We dedicate our toil.
To God and home and country,
As tillers of the soil.

Our roots are in these fertile vales
And in these rocky hills.
We graze our heifers in these dales
And fish the rushing kills.

Old memories are strong here;
Our fathers worked this land.
The ghosts of heroes throng here;
'Twas here they took their stand.

With firelocks at the ready,
They met the Crown's attack.
Their aim was sure and steady;
They drove the tyrant back.

Then bent they to the arders
And did not shirk the chores.
They filled our country's larders
And fought our country's wars.

And those who come behind them,
They hold this land in trust.
The ancient hills remind them:
Your dust is of our dust.

To God and home and country
We offer up our best.
In God and home and country,
Our future shall be blessed.

This volume is a celebration of the farm country of upstate New York, in and around the part of Washington County where Elizabeth grew up. It is a love letter to the land and its people. When Elizabeth and I visit her hometown of Greenwich, we are always drawn to the surrounding countryside: the wooded hills, the river valleys, and, most of all, the farms.

When Elizabeth and I had completed our first book, *Upstate New York: Towns That We Love*, and found it well received, we thought about what to write next. At first, we had the idea of Maryland port towns, since I am a Marylander and love the tidewater. But a chance conversation with Meg Southerland at Gardenworks Farm in Salem changed our plans completely. When we asked her what she thought we should add in a second printing of *Upstate New York*, she suggested including more of the farms, with pictures of the people and animals. She told us of friends and neighbors who were keeping up the farming tradition, and she and her husband showed us their own beautiful farm and store. We decided then and there that we needed to cover this topic, not as an addition to the first volume, but as a whole new book. Our publisher liked the idea and set us a six-month deadline. Although at first I was dismayed by the short time available for research, interviews, and writing, I soon realized that it was actually a blessing. There are so many different types of farming operations, so many delightful and inspiring people to meet, and, even within the small geographical area of interest to us, so many farms competing in the same lines of production (such as alpacas, sweet corn, milk, maple syrup, and so on). The short deadline made it manifestly impossible to do a systematic evaluation of the different producers. Of course, an exhaustive report was never our intent. Rather, our attention landed serendipitously on some farmers and not others, through fortuitous meetings and tips from friends. No one was intentionally left out.

Farming holds a special place in the human consciousness. The fact that farms produce the raw materials for most of our food and clothing ("No Farms, No Food," as the bumper sticker reminds us) is quite beside the point. Neat rows of crops, well-tended barns, and contented grazing animals all speak to our hearts with an intensity

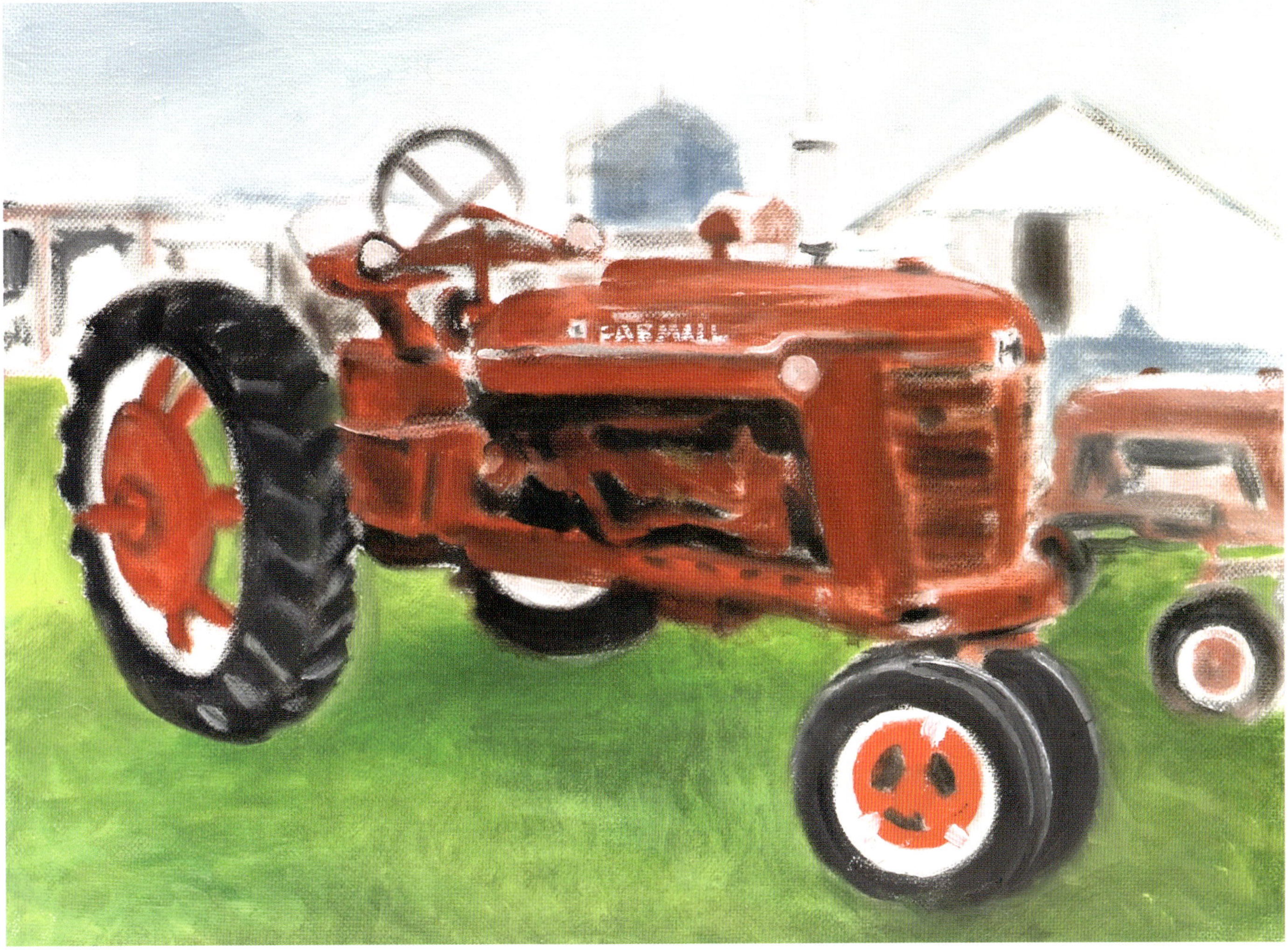

of feeling that cannot be explained by merely utilitarian considerations. With its rich evocations of home, hearth, and harvest, the very word "farm" has a feeling of security about it, as well it might, derived as it is from the Latin *firmare*, "to make firm." Even in these latter days, we can imagine what a sense of relief our ancestors felt as they looked out over cleared, tamed land, having hewn it out of the howling wilderness.

Yet the farm also reminds us of the fragility and transience of all human endeavors. Unworked fields revert to forests. Barns and houses do their best to fall into ruin. Drought, flood, lightning, wind, hail, frost, blights, and insects threaten the crops. Even a good harvest may be a disaster if it creates a supply that exceeds demand. And whatever nature has left standing, the government may destroy through some change in the rules: taxes, regulation, or deregulation—anything that upsets the plan. Or human innovation may render old ways obsolete. The spirit of the pioneer must animate the farmer even today.

The seasons, which set the rhythm of farm life, are at once an organizing principle and a metaphor. The life of the land is a cycle of renewal, day by day, season by season, generation by generation. Farming itself is in transition and has been since Stone Age people in Mesopotamia started it thousands of years ago. Working the land and

We visited an antique tractor show at the Washington County Fair, and I fell in love with this old 1950 Farmall (H series) tractor that was on display. It was hard to paint; this was my third attempt, because Barton said that if I didn't get it right, all of our farming friends would never let us live it down.

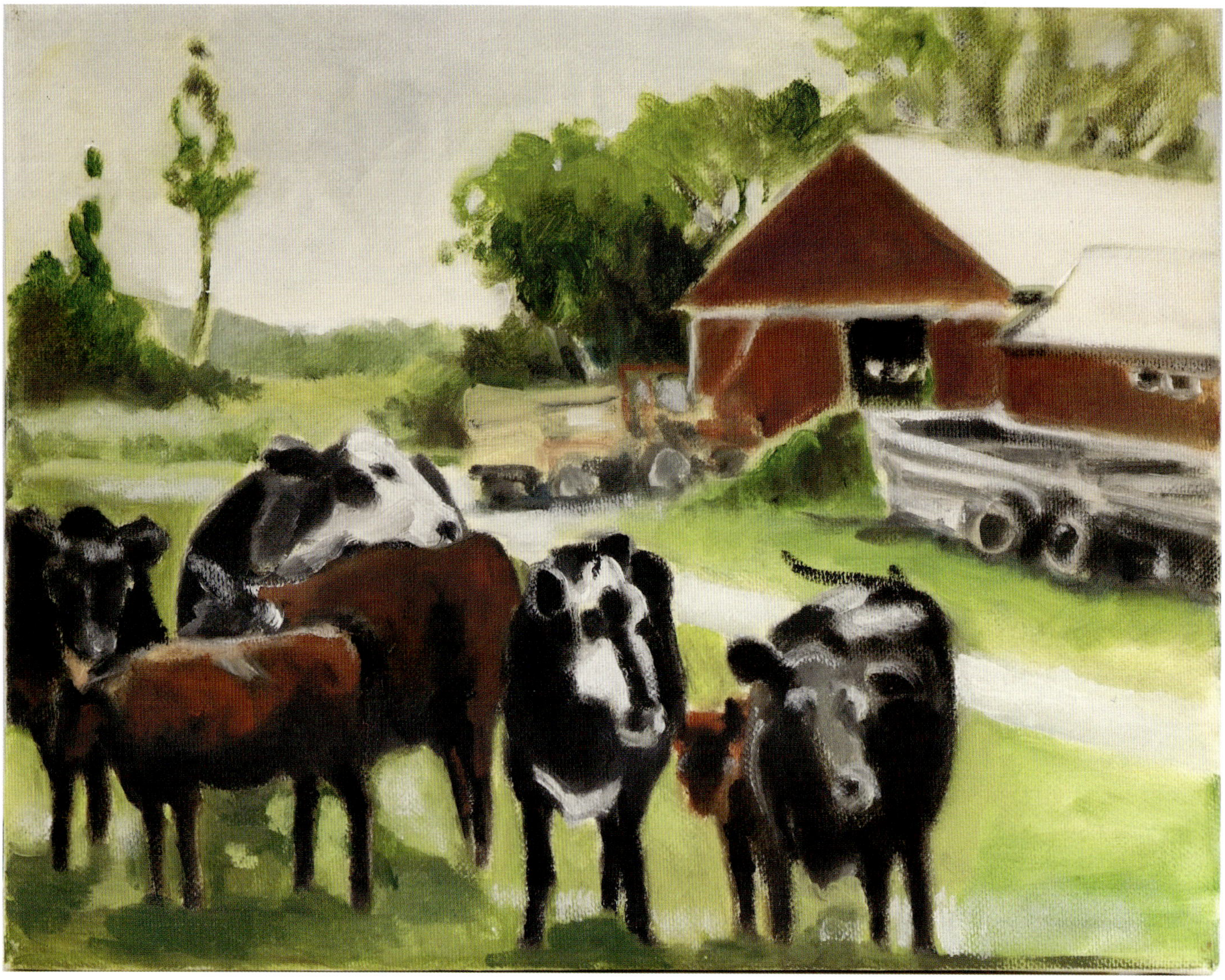

Cows seem to huddle in groups, probably as protection from predatory animals. These animals are owned by my good friends Glen and Jessica Townsend in Argyle, New York. They are surrounded by rolling pastures and beautiful vistas that look out toward the Adirondack Mountains.

making it pay continue to be a challenge. Specialization and bigness seem to be the trend. International competition further undercuts prices of farm products. The almost completely self-sufficient farm of the nineteenth century is no more, and the small family farm is said to be dying out. Yet some small farming enterprises are thriving, as their owners find new opportunities in the growing demand for organic food, locally grown produce, and specialty products.

The land in this part of New York has been under cultivation since before the Revolutionary War. With the end of hostilities between the colonists and the French in 1765, Indian attacks ceased along the frontier, and pioneer families began to clear the land and plant crops. The respite was brief. In 1777, General John Burgoyne, with British regulars, German mercenaries, and thousands of Indian warriors from the frontier, terrorized the region for a season. The British defeat at Saratoga removed the most pressing danger, but life did not return to normal until the close of the war, when sporadic raids by Tories and their Mohawk allies came to an end.

The planters who took up arms to bring about the birth of the American republic were risking all that they had labored to acquire. Schooled in the stories of classical antiquity, the officers of George Washington's army looked to the example of the Roman patrician

Lucius Quinctius, who was called Cincinnatus because of his curly hair. Cincinnatus, at the request of the Roman senate, assumed the title of dictator in 458 BC and again in 439 BC to lead his countrymen against the invading barbarians. He left his farm, took up his sword, and did what needed to be done. On each occasion, the enemy having been defeated, he relinquished his power and returned to his farm. Eighteenth-century paintings of the hero show him handing back the *fasces* (a bundle of sticks with an axe blade, symbolizing strength in unity) to a female figure representing the Roman republic.

This powerful image, linking heroic virtue with the renunciation of tyranny, is one of the central icons of our nation's founding. I think there is a basic modesty that comes naturally to the farmer, a recognition of the essential order of things. You can't make a plant grow any faster by issuing a court order or passing a legal statute. Cows behave like cows, and pigs behave like pigs. Abstract theories don't last long unless they accord with practical realities. Farmers have enough to keep them busy without minding other people's affairs, and have too much sense in any case to want to order other people around. It is no wonder that Thomas Jefferson considered an agrarian economy to be the font of liberty and the reservoir of public virtue.

As the heirs of this noble tradition face the uncertainties of the twenty-first century, new challenges bring new opportunities. In Washington County and the nearby parts of upstate New York, much of the agricultural activity used to focus on dairy farming. But with competition from industrial agricultural methods and supply outpacing demand, milk prices have dropped, and formerly prosperous farms have fallen on hard times. Elizabeth remembers that in 1956, when she was little, she went with her grandfather, Guy Barber, to a milk-processing plant, where he gave a speech to the other dairymen assembled there and persuaded them to dump their milk rather than accept the prevailing price. Guy emptied his milk cans, and the other farmers did the same. After that, her grandfather told her, "It's all over for the small dairy farmer." He sold his herd, and today what was once the calf pasture is a forest.

Yet others have stayed on. Our friend Glen Townsend and his wife, Jessica, own and operate a small dairy farm with a herd of forty cows. Jessica, whose family has milked cows since the eighteenth century, says this kind of farming is in her blood and that she feels an attachment to the life of the dairy farmer. "We have a small herd, and I know every cow by name," she said. "They are like members of the family."

Glen loves the cows too, but he is more practical in his approach. "At this point, every day I'm doing this, I'm losing money," he said. "It costs at least eighteen dollars per hundredweight to produce the milk, and it's selling at twelve dollars. Dairy farmers are borrowing money to stay in business. In this economy, you can't make it up on volume. The big operations are borrowing more, but they have the greater potential for profit when things turn around."

This arrangement sounds suspiciously like that of a highly-leveraged commodities speculator, but with the added benefits of fifteen-hour workdays and manure on your boots.

Teri Ptacek, the executive director of the Agricultural Stewardship Association, is quoted in the Glens Falls *Post-Star* as saying, "If prices remain like this, we're not going to have dairy farms."[1] The association helps farmers sell the development rights on their property to the state of New York, so that the land remains dedicated to agriculture. But even if a dairyman receives a cash payment, he is still in the position of the farmer in the old joke, who, when asked by a reporter what he's going to do with the million dollars he won in the lottery, replies, "I guess I'll just go on farming 'til the money runs out."

Yet even in this atmosphere of anguish and uncertainty, there are many rays of hope. And one thing is certain: the indomitable spirit of the Yankee farmer will prevail.

1. Reisman, "Fearing prosperity Farmers worry."

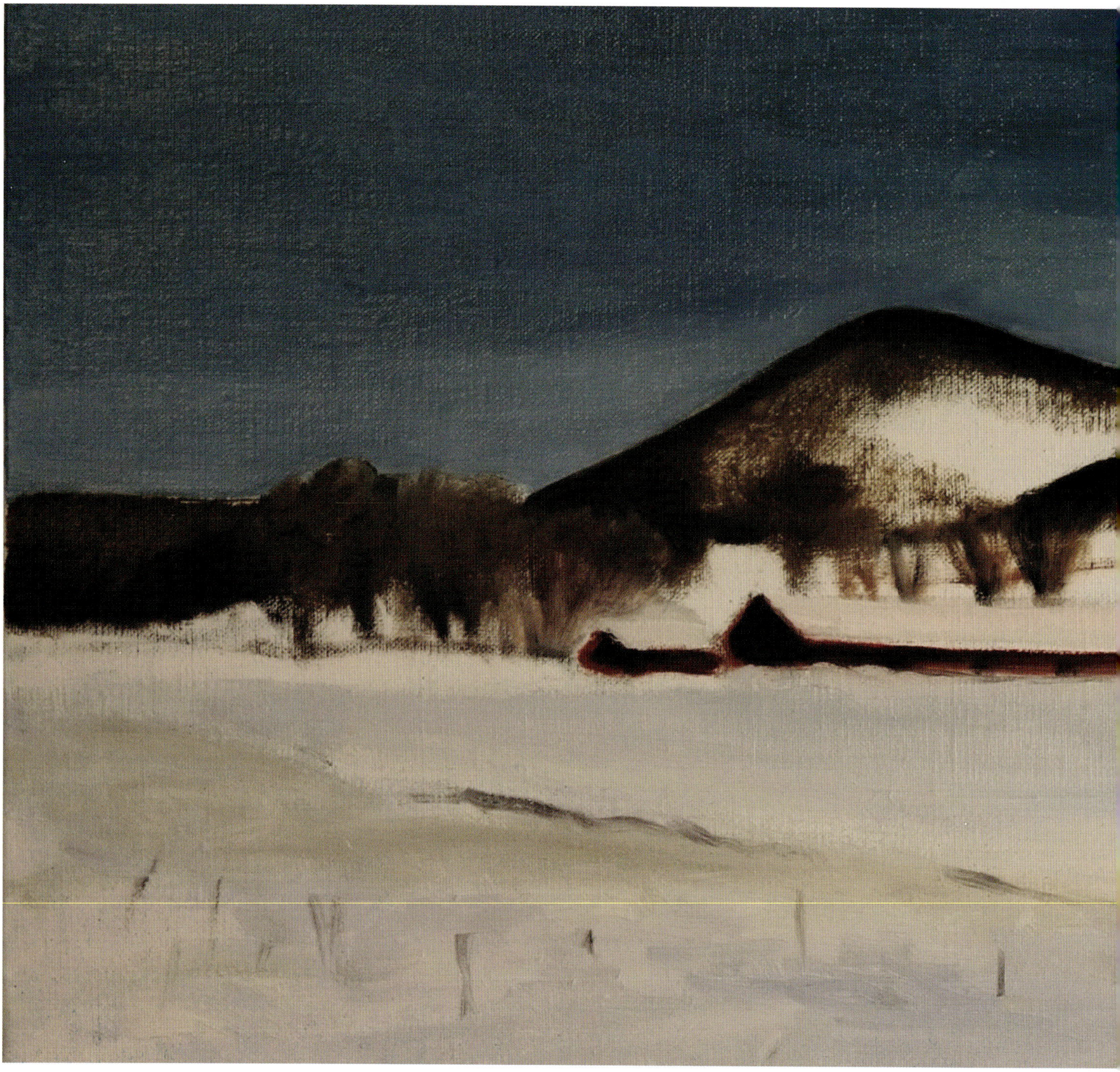

WINTER

Winter and discontent have occupied the same space in our minds as a species since long before Shakespeare immortalized the connection in his *Richard III.* For one thing, it's cold. And in upstate New York, it's *really* cold. But the hardy settlers and their successors made the most of the low temperatures. They cut blocks of ice from frozen ponds and rivers and packed them in straw in their icehouses: refrigeration without electricity. They built the famous Yankee clippers, which sailed so fast that they could sell ice in India.

As an alternative to distillation, they froze fermented apple cider to concentrate the alcohol, creating applejack. For those with less patience, there was a drink called a "Stonewall," favored by Ethan Allen and the Green Mountain Boys, made from hard cider and rum, so called because its consumption eased the construction of all those stone walls on the farms. Elizabeth and I are

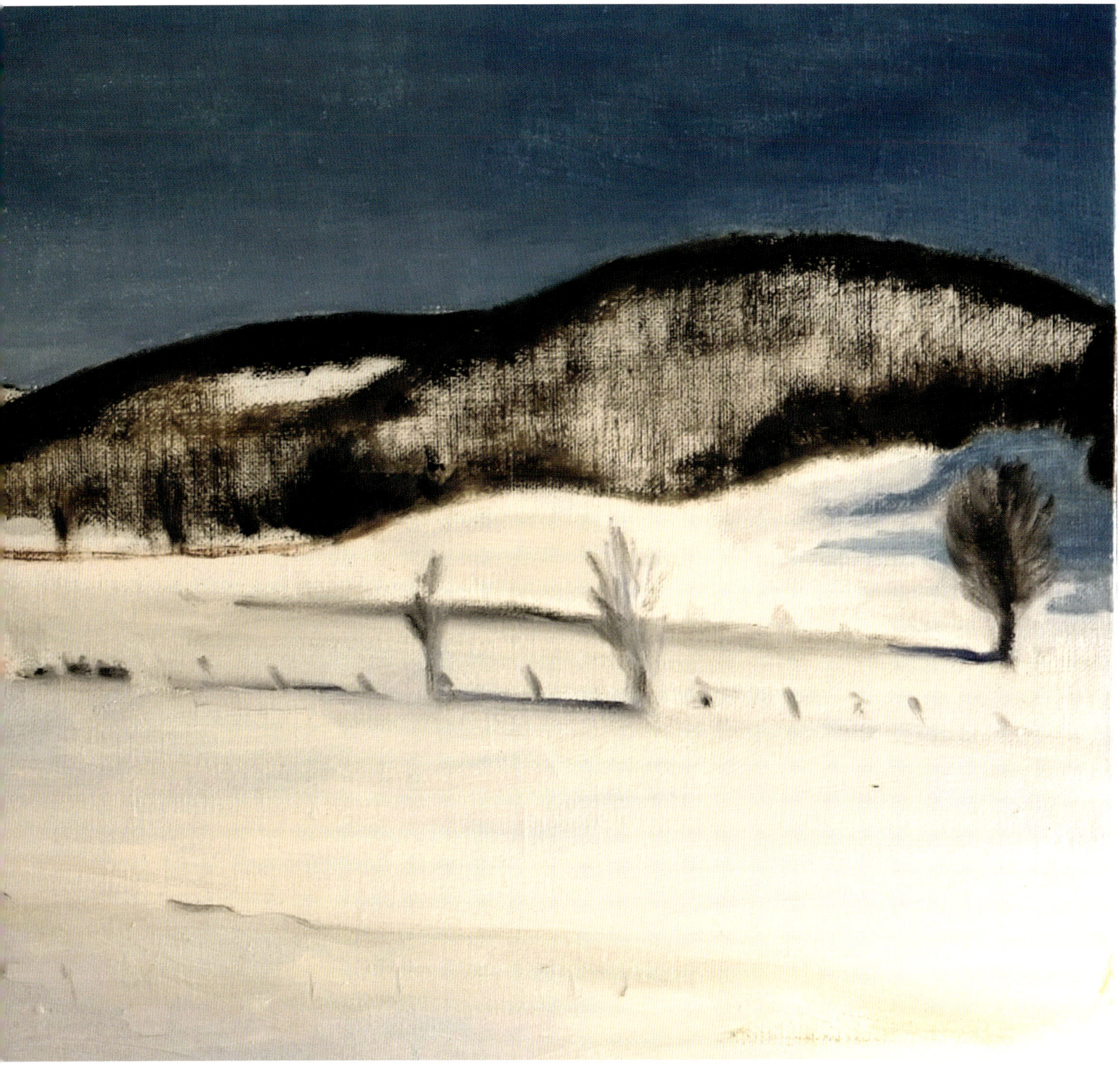

particularly fond of the cider that comes from New York McIntosh apples, but it doesn't go hard the way fresh cider used to do. The sterilization process, required by law to prevent the spread of salmonella, kills the natural yeasts and bacteria too. We'll just have to save up for our own cider press.

The sap is down in the winter, and felling trees for lumber and firewood keeps a man warm. There are few activities more deeply satisfying than splitting logs. Well, maybe that's a minority opinion, but *I* like to split logs,

Chambers Valley is nestled in the hills of New York and Vermont. The surrounding area is used solely as farmland and, of course, for maple syrup production. Located between the historic villages of Salem, NY, and Manchester, VT, Dry Brook Sugar House is a great place to visit in the early spring during boiling season. This painting was based on one of Jon McClellan's beautiful photographs.

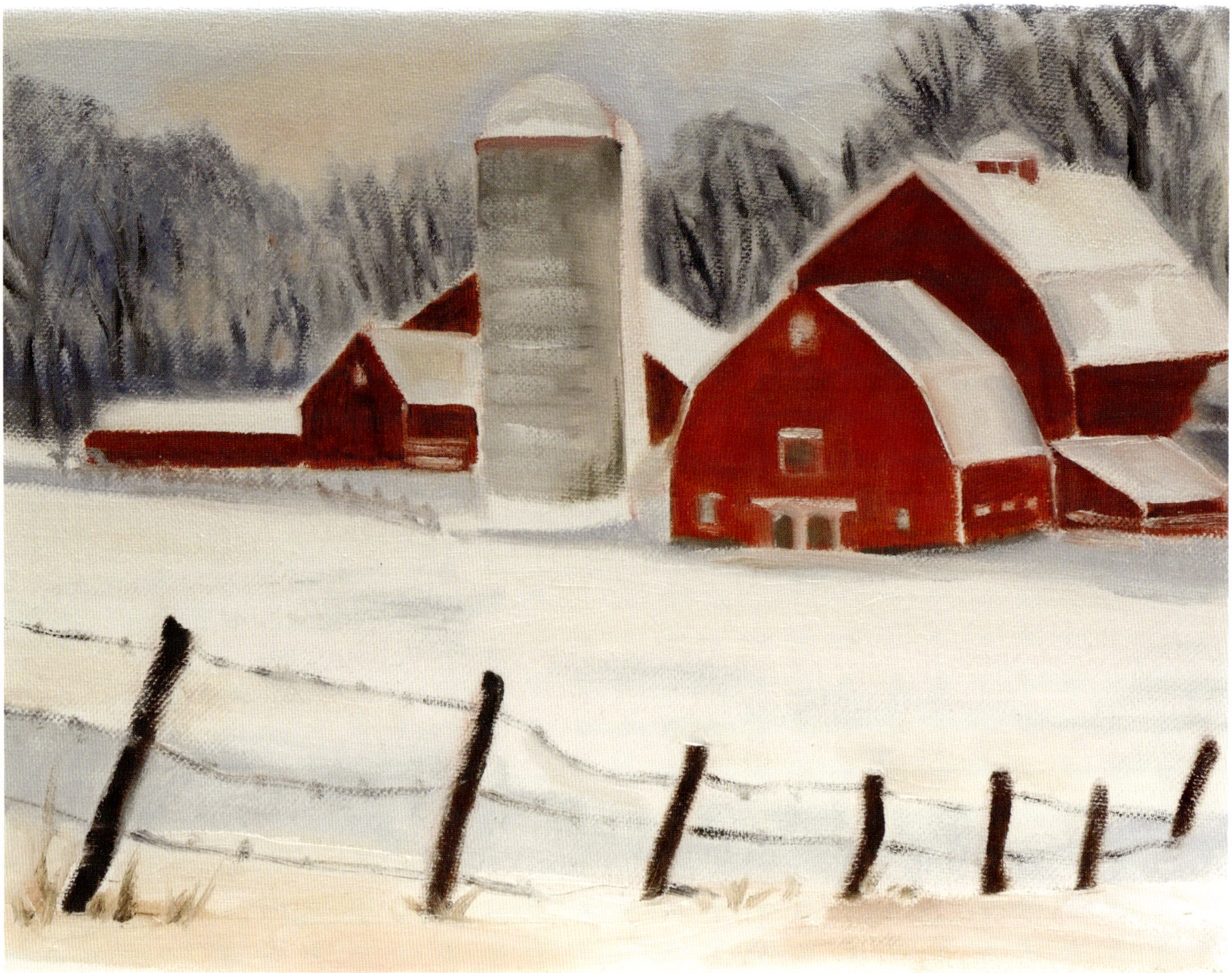

We do not travel to our hometown of Greenwich much during the winter months; perhaps our blood has thinned over the years, or maybe we are just too old to stand the icy conditions. Fortunately, my good friend Bob Olbrycht took pity on me when I told him I was in desperate need of winter images from which to paint. He e-mailed me his photograph, which I used for this painting of barns in winter.

and not with a newfangled hydraulic splitter, but with an old-fashioned splitting maul. If the grain is too contrary, I use wedges, but that feels like cheating. The grain of wood is a model and metaphor for everything in life: there is a grain to reality, and we succeed best when we pay attention to it. The earth will bless us with her abundance if we pause to pay attention to how things work. The speed of light, Planck's constant, the way the fibers run in a tree crotch: is any of it a good idea? I don't know, but it's the law.

There is a reason for the thrifty character of northern peoples: if you don't save up food, fuel, and fodder for the cold months, you may remove your family from the gene pool. The same principle applies to money. The old farmers in these parts distrusted banks. They were capable of holding onto their own money, and when they found it necessary to part with some of it, they would peel the bills off of a big roll that they carried with them. Those who have continued to live frugally and avoid debt have fared the best in the most recent market downturn.

The Hendersons are such a family. Farming the same land in Easton since the early 1900s, they have remained in the dairy business through good times and bad. Elizabeth grew up with Alan Henderson. She still calls him by his nickname, "Mister," and he still calls her "Betsy," like all of her old friends in Greenwich. They both still call his younger brother "Squirt." Alan showed us his barn, milking parlor, and calf pens. As we walked through, we met Squirt's son and daughter with the prize cows they were grooming for the county fair. These two cows, like all of their animals, were beautiful and well husbanded. The milking parlor was clean and well maintained.

With the quiet competence and brilliant modesty of the successful farmer, Alan explained that his family has held on because they all help with the work. His mother milks the cows every morning at four, and his father gets an early start too. A thin man with a twinkle in his eye, ninety-year-old Taylor Henderson has more energy than most men half his age. Shifting from one foot to another as if he were hearing music and looking for a dance partner, Taylor explained that his job on the farm is to get into trouble. Well, somebody's got to do it. He also drives the tractor, sowing and harvesting corn, wheat, and soybeans.

The Hendersons know how to repair their own equipment, and they have not gone into debt to buy shiny new tractors and milking machines when the old ones work just fine. They still use their silos, and they mix their own feed from crops grown on their own land. Working over seven hundred acres, they keep a herd of about one hundred ten milking cattle, and have resisted the temptation to expand beyond a workable size. Teamwork, frugality, and multigenerational farming experience are their formula for success. May it long continue.

When the leaves are off the trees, it is easier to see the farm buildings. Many of the structures in Washington County and its environs are from the late eighteenth and early nineteenth centuries, of sturdy post-and-beam construction, with roofs of local slate. Barns and other outbuildings define the farm landscape and add human scale to the vastness of the fields. Icehouses, corncribs, even a few privies still stand on many of the old farms. Although no longer used for their original purposes, the old buildings have stayed put for a variety of reasons, not least of which is the fact that they were crafted with a keen aesthetic sense. An economist would say they add value as improvements to the lot. I think the improvement goes beyond any utilitarian calculation and speaks to our basic need for order and structure in our surroundings. The old outbuildings are a comforting presence, a reminder of bygone ways of doing things, and a tangible legacy of those who planned and dreamed and built.

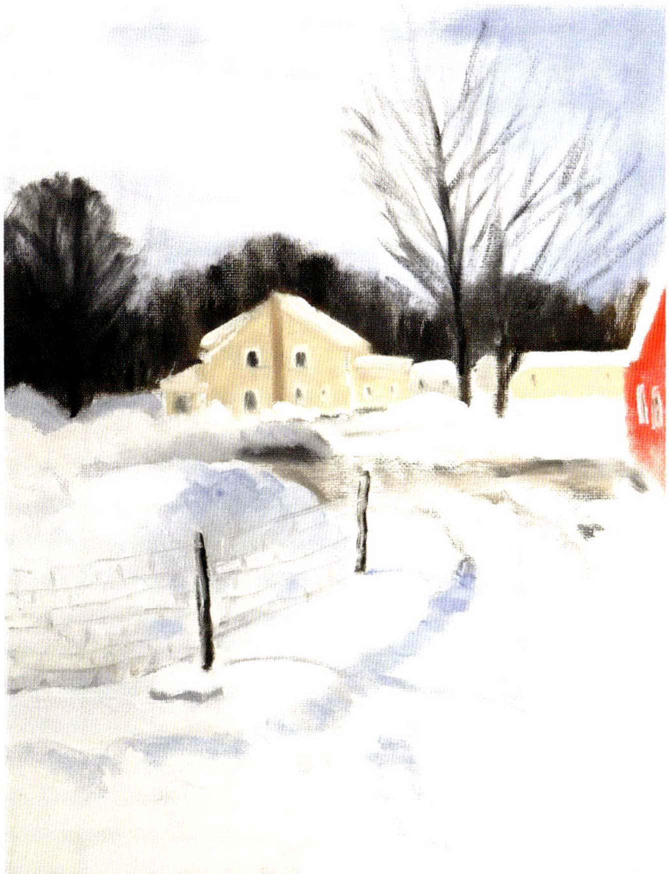

I think this painting captures how lonely it can be in the dead of winter, when you can hear and feel your feet crunch on the icy, hard-packed snow. The temperature sometimes gets to subzero in these parts. This house and barn are part of the original Jablonski family's Argyle homestead, which they have been farming since the early 1900s.

At Gallery 668 in Battenville, the barns that used to be full of cows and hay are now full of art. Lovingly restored by Solange Batsell Herter and improved and maintained by her daughter Veronique de La Bruyere, the nineteenth-century structures are a perfect venue for artists, musicians, and patrons of the arts to gather for concerts, exhibitions, and gala receptions. The work of contemporary artists comes and goes with the seasons, but the craftsmanship of old-time builders remains on permanent display in the massive, exposed, hand-hewn posts and beams, with their mortises, tenons, and trunnels. An

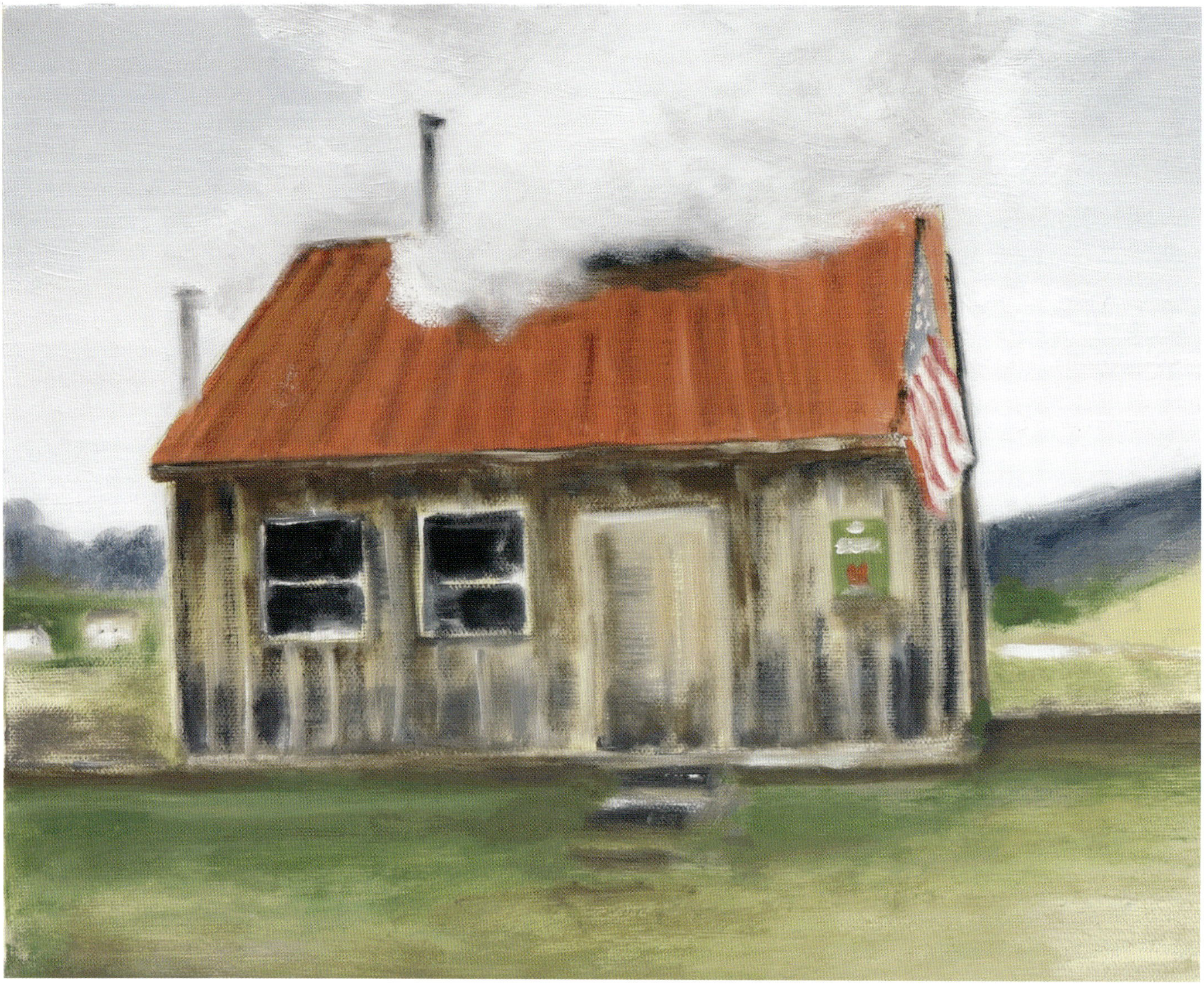

old round wooden silo stands near the garden path, and an arched passage frames a vista of green lawn leading to the rambling white frame house that was once the Battenkill Inn. Built in 1779, the inn is remembered for its service as a stop on the Underground Railroad and as a place where nineteenth-century political activist Susan B. Anthony stayed. Today it is the home of Veronique de La Bruyere, and its decor reflects her sophisticated taste and her interest in local history. Schooled abroad, and traveling widely with her family, she loved to come here on vacations. Elizabeth asked her why, with all her experience of fabulous, exotic places, she chose to live here. Veronique answered simply, "Because this is the most beautiful place in the world."

Obviously, Elizabeth and I share Veronique's opinion. One of the pleasures of painting a landscape is that one notices details that are either too big or too small to retain the attention of casual observers. Silos are such a familiar feature of the rural landscape that we tend to take them for granted, but if you could go back to when the boys marched off to join the Union army, there would not be a silo in sight. Stored in barns or stacked in fields, hay provided food for the cattle during

John Reid's family sells maple syrup from his sugar shack, located right behind the farmhouse and cow barns, just east of Greenwich, New York. The day we visited their farm, I ran into one of my earliest childhood friends, Debbie (Dewey) Henley. It's great going home and seeing people you grew up with. Debbie and I used to play up the road from the Reid farm at her parents' place. I think all of us lived on dairy farms back then.

the winter. Back then, the idea was simply to keep the cows alive, not to maintain the high level of nutrition needed for year-round commercial milk production. It is no mere accident that the growth of dairy farming as an industry coincided with the development of the silo.

Storage of animal feed has always presented a challenge to the farmer. The ancient Syrians stored feed in pits in the ground, and the Greeks used the word *siros* for their airtight grain pits, hence the word silo. In 1875, the U.S. Department of Agriculture published an article entitled "The French Mode of Curing Forage," which introduced the basic concept of silos to American farmers.[2] One of the earliest applications of the principle was a pit in the ground, six by thirty feet, and six feet deep, filled with alternating layers of rye straw and chopped corn stalks, made by Levi P. Gilbert of Fort Atkinson, Wisconsin, in 1877.[3] Now ground storage is back in vogue, with the attractive addition of black plastic and old tires to cover the silage. Holes are cheaper and easier to maintain than towers. Consequently, although many of the picturesque silos in our landscape are still in use, many are not. Those that still stand are generally of durable materials and will last well beyond our lifetimes, even if they all fall into desuetude. The great age of the silo appears to be ending, little more than a century after it began.

In 1882, Vicomte Arthur de Chezelles attended the Royal Agricultural Show at Reading, England, and told an international audience about his thousand-ton-capacity silo that could hold hundreds of acres of forage. Soon, farmers in both Europe and America were experimenting with silo construction. Most of the early ones in the 1890s were square wooden towers, which evolved into round ones to simplify construction and eliminate corners where air pockets led to decay of the silage. The early twentieth century saw the development of concrete staves that fit together into a perfect cylinder, reinforced by iron hoops like a barrel. Ceramic tiles and poured concrete also came into use. The grandest silo of all, the Harvestore, recognizable by its smooth, dark blue exterior, came out of the beer industry in the late 1940s. Made by the A. O. Smith Company of Milwaukee and inspired by beer vat design, this glass-lined steel silo covered with blue fiberglass was said to mark a farm as either rich or in debt.

The principle of the silo is to create an airtight vessel in which the chopped-up forage will ferment just enough to use up all the oxygen and make enough acid to prevent decay of the cattle feed. The creation of silage takes three weeks and is divided into five phases. In Phase I (days zero through two), the still-living plant cells in the chopped-up forage and the aerobic bacteria riding along find themselves entombed and begin a short spell of respiration, producing CO_2, water, and heat. The plant cells die, oxygen is depleted, and Phase II (days two through three) begins as the bacteria change over to anaerobic respiration, producing acetic acid. The temperature inside the silo reaches a maximum, around ninety-five degrees Fahrenheit. During Phase III (days three through four), acetic acid production continues, but the bacteria also start making lactic acid. In Phase IV (days four through twenty-one), lactic acid production continues, and the temperature and pH drop. After twenty-one days, during Phase V, the silage is done, pickled with acetic and lactic acid, which are themselves nutrients for cattle. It will keep all winter, as long as it is not exposed to oxygen.

Winemaking it ain't, but there is still a fair amount of finesse required to make good-quality silage. The most critical elements are moisture content and available sugars. If moisture is over 70 percent, clostridia (bacteria in the same family as those that cause the diseases botulism and pseudomembranous colitis) will grow, making butyric acid and sour silage. Too little water, and the silage overheats, cooking the proteins and reducing the availability of nutrients. If there is not enough sugar, acid production will be insufficient to lower the pH to the desired range of 3.8–4.2. Corn is easier to ensile than hay because it contains more sugar. And size does matter: silage chopped too fine leads to digestive problems, and overly coarse chunks trap air pockets that interfere with fermentation. There is a lot to know, and it is easy to see why it helps if your family has been building up experience for several generations.

2. Shaw, *Soiling Crops and the Silo*, 256.
3. Current, Wisconsin: A History, 75–6. The exact size and date of Levi Gilbert's silo vary with the source.

SPRING

The first sign of spring is the rising of the sap in the sugar maple trees. In the beginning of April, we visited Sugar Hill Farm to buy some fresh maple syrup. There was a sign on the road, but there was no mistaking the plume of steam rising from the boiler. It was a chilly day, but inside the sugarhouse, it was warm and steamy. This visit was a treat for me, as I had never seen maple syrup made. Growing up in Baltimore, I had come to regard real maple syrup as a rare luxury, to be bought in a small bottle and applied sparingly. Yet here were more choices of syrup than I knew existed: light, amber, and dark amber. They all tasted good, but the darkest syrup was the best, with the richest, most concentrated flavor. My favorite product was maple butter, a spreadable form of maple sugar like honey butter, absolutely addictive on toast, and not excessively sweet.

Vermont is better known than New York for maple syrup, not just because of that cute Vermont Maid on the bottle, but because it produces more: 920,000 gallons in 2009. In the same year, New York turned out 362,000 gallons, behind Maine's 395,000. Canada makes the most maple syrup, though—about 85 percent of the world's production. Syrup production is a good supplement to the income of a small farm, as the price is high and a farmer can sell directly to the public at retail rates. The weakening dollar has pushed up Canadian syrup prices, and New York syrup has risen in price accordingly.

The method of collecting sap is less picturesque that it once was. The classic galvanized steel buckets have been replaced by miles of plastic tubing and large plastic storage tanks, often with vacuum pumps to improve the flow. It takes a lot of sap to make syrup. You can calculate how much sap is needed by using the "rule of eighty-six." Measure the sugar content of your sap, and divide the percentage into eighty-six. If the sap is 2 percent sugar, you'd divide eighty-six by two for a result of forty-three gallons of sap to make one gallon of syrup. As a general rule, one tap yields one-third gallon of syrup.[4] To tap a tree, the farmer drills a 7/16-inch-diameter hole, angled slightly downward, inserts a metal or plastic spile, or spout, then gently seats the spile with a hammer to create a tight seal. Plastic spiles work best with plastic tubing.

Tapping does not harm the trees, as long as the number of taps is kept to a minimum. The current recommendation is one tap for a tree twelve to eighteen inches in diameter, two for eighteen- to-twenty-five-inch trees, and three for larger trees. Thus, the syrup maker maintains a healthy woodlot, termed a "sugar bush," which usually contains a mix of hardwood trees, such as beech, ash, and yellow birch, in addition to maples. Judicious harvesting of the valuable hardwoods allows the tree crowns to spread and adds a bit of income.

Once the sap is collected, the evaporation process begins. Old-time copper kettles over wood fires have given way to commercial evaporators, usually heated by oil. No one seems to be pining for the old ways. Modern equipment produces syrup of consistently high quality,

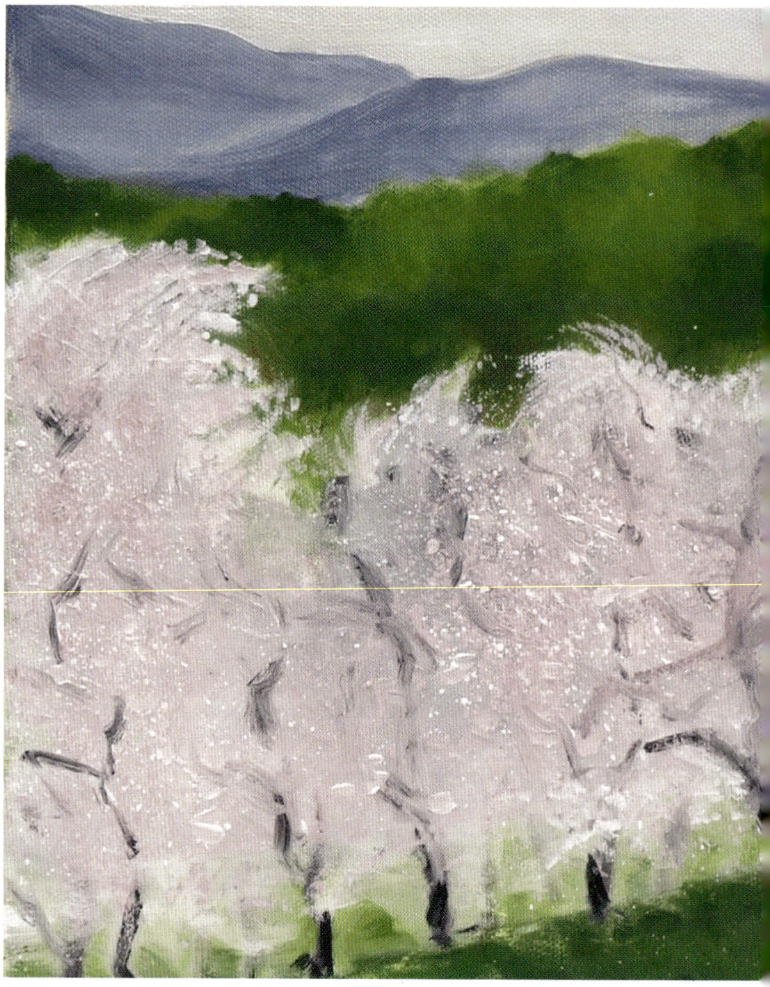

4. Cornell Sugar Maple Research & Extension Program, "Frequently asked questions."

and you can still spot a sugarhouse from a long way off by the plume of white vapor that it sends up into the chilly air.

As the buds were opening and the fruit trees blooming, Elizabeth and I made one of our many visits to the Saratoga Apple orchard in Schuylerville. The view from the hilltop over the orchard to the Saratoga Battle Monument and the hills across the Hudson is always beautiful. It was one of those dramatic afternoons in May, when the rain has washed everything clean, and the rays of the declining sun have burst through to light up the apple blossoms like fireworks against the leaden storm clouds to the east. Innumerable bees hummed among the flowers. Before European settlers came to North America, our continent had no apple trees. All these trees are the carefully selected and grafted descendants of wild apple trees that grew in the Caucasus thousands of years ago. And these golden brown honeybees, the world's greatest pollinators, without which the trees would bear no fruit, also came over with our colonial forebears.

The first honeybees in the New World are said to have been brought over in 1622. The American Indians called these novel insects the "white man's flies." Of course, the bees swarmed out of their hives and took off across the country, stopping when they reached the Rocky Mountains. Early settlers in Texas discovered huge bee caves full of honey, little suspecting that the bees had arrived there only a short while before they had.

This apple orchard is in full bloom, a striking contrast to the Vermont hills that lie in the background. The scene lies southwest of the town of Schuylerville off Route 29. On the other side of the road is a farm stand owned and operated by Saratoga Apple. It's open all year long and boasts a variety of local apples, cider, fresh or frozen vegetables, jams and jellies, woven baskets, and fresh doughnuts every day.

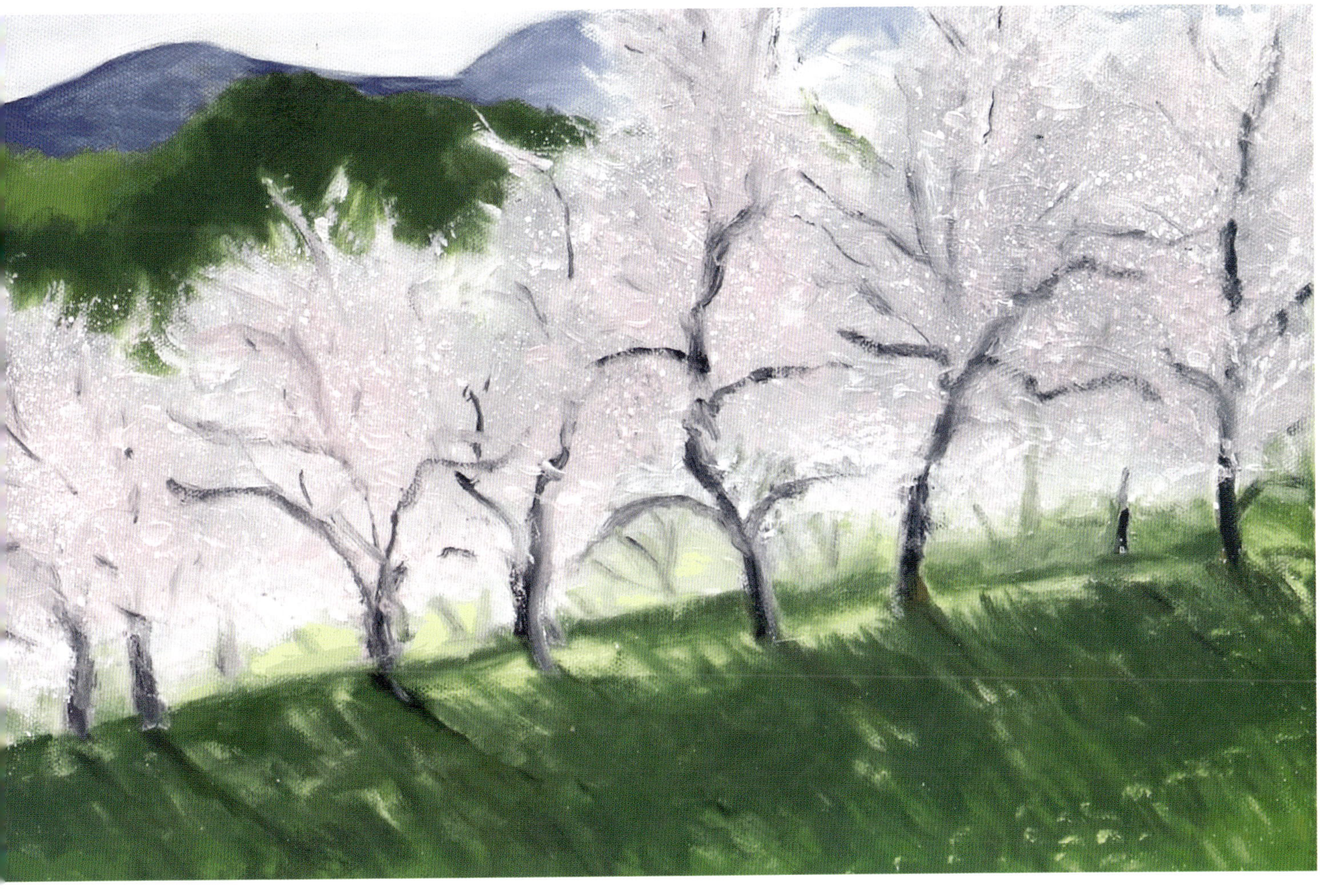

Despite their success as an introduced species, honeybees do best under the care of humans, having thrived that way since before the Egyptians built the pyramids. Evidence from rock paintings suggests that people were keeping bees as long ago as 13,000 BC. Today, kept bees are in great demand, not only as producers of delicious honey but also as pollinators of our crops. Americans consume hundreds of millions of pounds of honey every year (about 1.1 pound per capita), and the commercial value of bee rentals for crop pollination exceeds fifteen billion dollars a year. In fact, the U.S. Department of Agriculture (USDA) estimates that 68 percent of beekeepers' income is from bee rentals.[5]

With some 35 percent of our food production depending on these little critters, the maintenance of bee populations is critical. But in the 1980s, those populations began a sudden decline. Two types of parasitic mites, imported by accident, began to attack the bees, and whole bee colonies began mysteriously dying off. The new plague, called colony collapse disorder (CCD), killed 30 to 90 percent of commercial bee colonies in over twenty-two states. The problem became so severe that in 2005, bees had to be imported from outside the U.S. for the first time in more than eighty years. In September 2007, researchers reported a link between CCD and a bee virus called Israeli acute paralysis virus. On the positive side, the scarcity of bees has pushed the price of honey higher than it has ever been, to $1.41 a pound in 2008, despite pressure from cheap imported honey.[6]

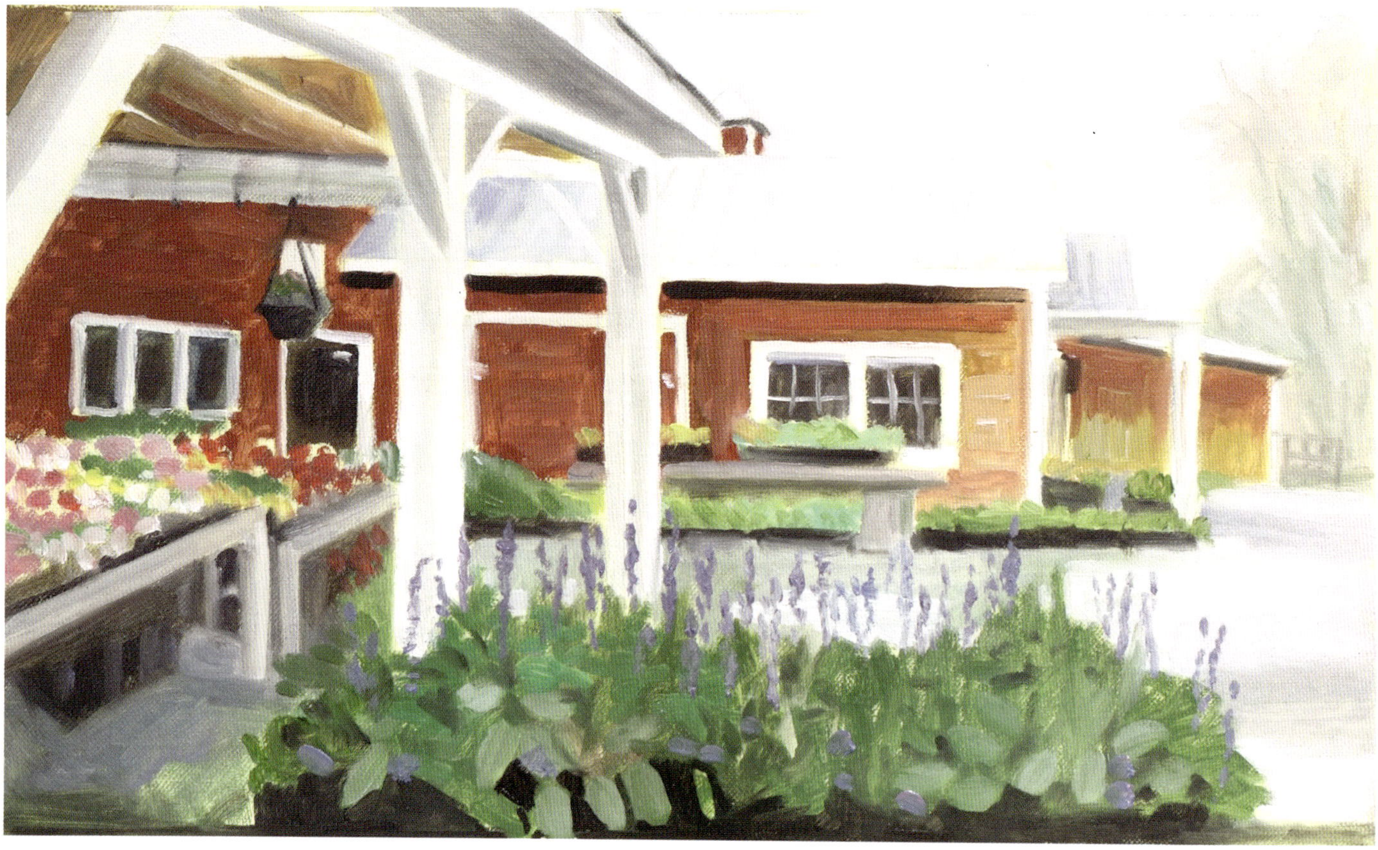

Gardenworks is a working farm where you can spend an entire morning or afternoon browsing around. The second floor in the main building has antique farming equipment on display. This painting shows the numerous outbuildings and barns that are part of the charm of the place. I took the photograph in early April, right after they had brought the early seedlings, lavender, and other hardy plants outside.

Greenwich, New York, is fortunate to have a nationally known bee business right on the western edge of the village. Betterbee was founded in 1979 by Bob Stevens

5. Agricultural Marketing Resource Center, "Bees profile."

6. Ibid.

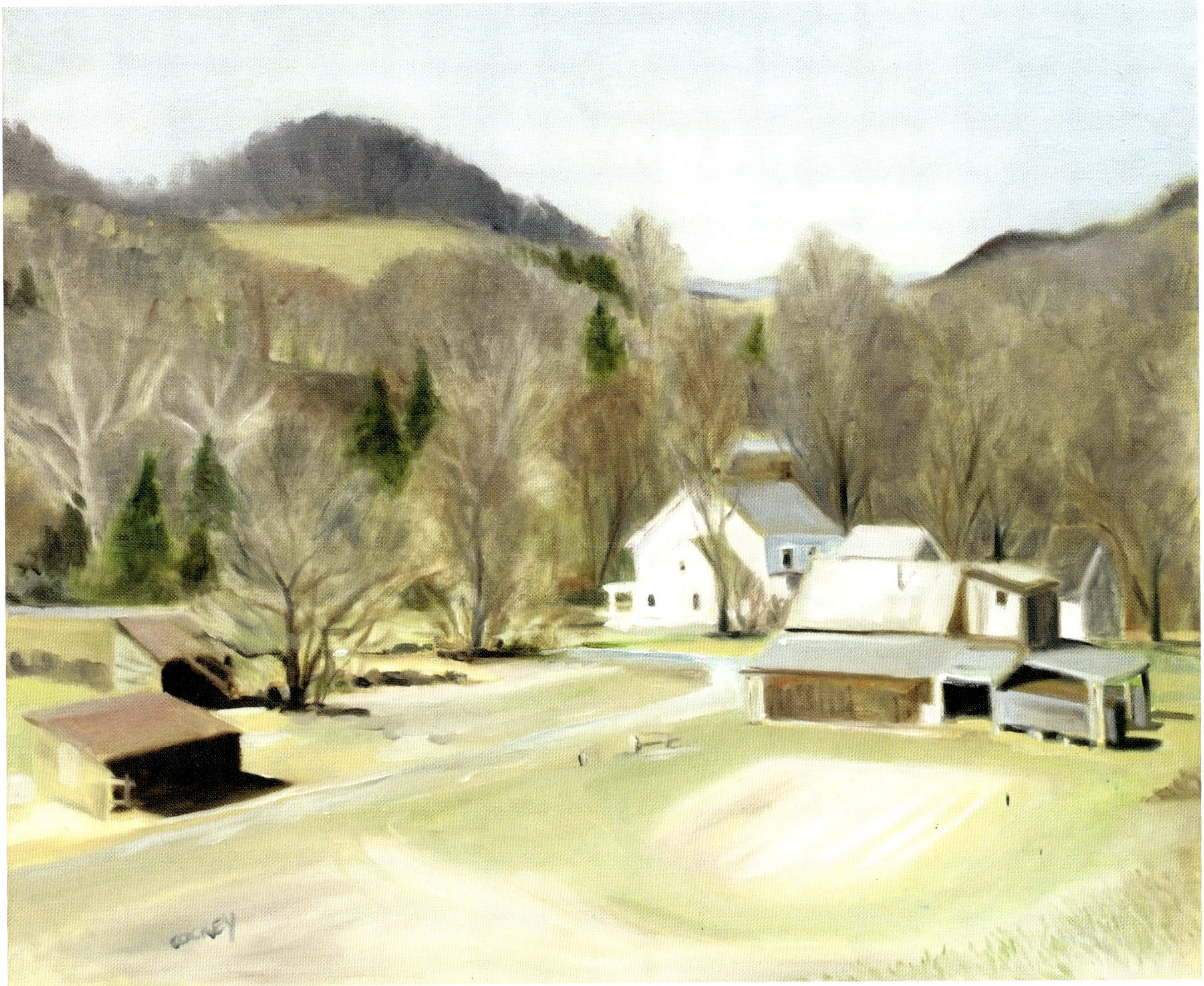

I like to walk up the hill behind my grandmother's house for a view of the Battenkill River, more noticeable in the spring when most of the leaves are still off the trees. Then again, Barton and I often visit this place in all seasons. I think we go back there just to make certain that the house has survived another season, that nothing has fallen down and the stone walls are still in place.

and his wife Margaret. Six years earlier, they had taken up beekeeping as a hobby to improve pollination in their gardens. Since then, the company has expanded to Internet and catalog sales, emphasizing its active involvement in the field with the motto "Beekeepers serving beekeepers." Located on the family farm, this thriving business sells everything needed for amateur or professional beekeeping, from the bees themselves to hives, smokers, protective clothing, and all the equipment for processing honey and wax. Elizabeth loves to mold Christmas ornaments from beeswax, and we bought an old-time Santa Claus mold at the Betterbee store. Do not visit Betterbee unless you are prepared to buy honey; it is simply too good to resist. Their honey is marketed under the label of "Southern Adirondack Honey." Most of it is produced on the farm, and the rest comes from local beekeepers.

I love the way the Stevenses describe the world of their bees. "Our bees work the wildflowers on the hallowed grounds where the battles of Saratoga were fought," they write on their web site. "They work the shores of the mighty Hudson River. They work the basswood trees along the world famous trout streams, the Battenkill and the Hoosick. They pollinate the apple orchards and the melon fields."[7] What could be better than that?

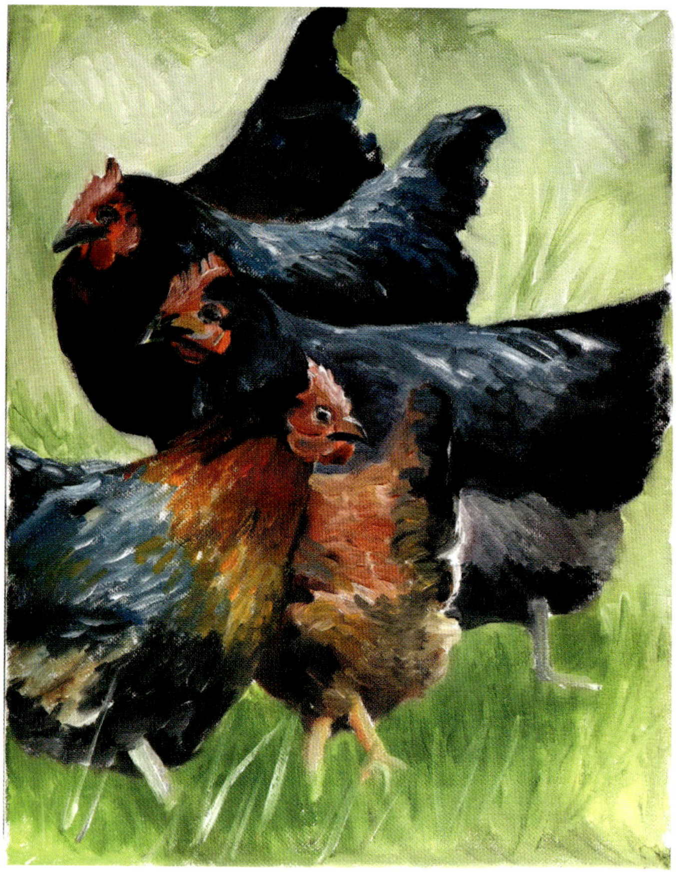

Driving by the Betterbee store one cool May morning, we had to slow down for a well-driller's truck. It reminded us that every farmstead needs a water source. Unless there is a reliable spring on the property, that source has to be a well. Although well drillers generally seem to be able to find an aquifer by drilling deep enough, many people hedge their bets by dowsing, or as they say in these parts, "water witching."

Below: The horses are kept in a beautiful pasture off the North Greenwich Road. They actually came up to the fence to greet me and check out my camera. One of the reasons Barton and I go home so often is because it seems miraculous that it's possible to visit working farms located only a short distance away from where we live in the village.

7. Betterbee, "Honey and Candles."

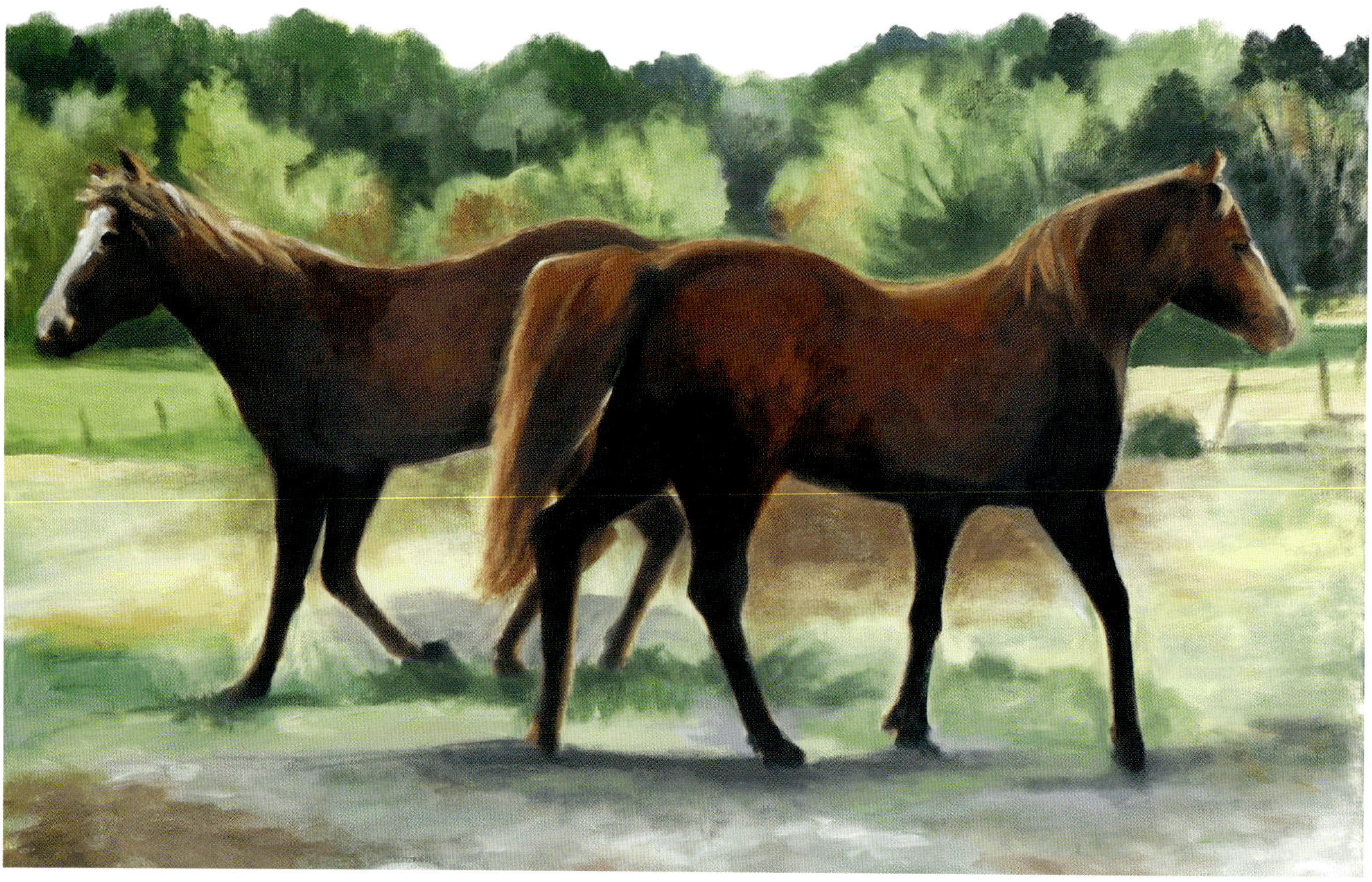

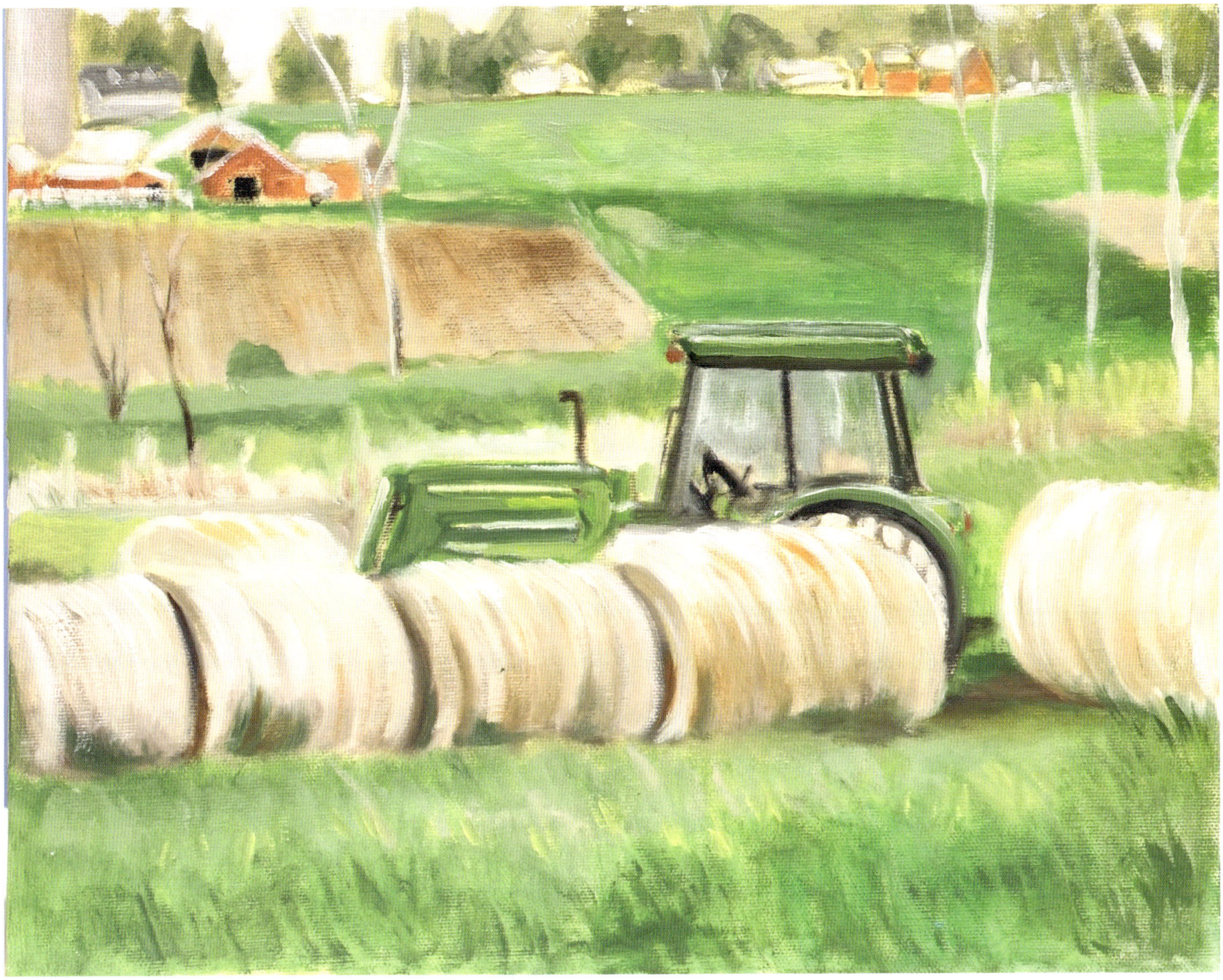

Opposite: The first time I saw these chickens, they were lying next to one another all over the ground with their wings spread out like fans. I was delighted and surprised, because I had never seen chickens do that. Bridget Cavna just smiled when I asked her what they were doing. "Sunning themselves," she said. "I guess you've never seen happy chickens before!" Indeed I had not.

Above: There used to be an airstrip where the green field is now. Our friend Charlie Tracy told us that he and Billy Becker used to land their planes there, right in the middle of the pasture. The bales shown here probably wintered in this spot. Shortly after I took the photo, someone came along and moved them all across the road.

My idea of dowsing always involved holding a forked stick, which was supposed to deflect when it passed over a source. As a child, I tried walking around with such a stick, but it never twitched. I concluded that either there was nothing to it or I had the wrong kind of stick. Water witchers today prefer metal tools. Some swear by welding rods bent at right angles. Our friend Bob Wright relates that his father could find water with an opened-up metal coat hanger and used to give demonstrations at the Washington County fair. George Perkins, who worked for the phone company, says that water witching is not specific to water. One day, when the telephone service for a large portion of the town went out of order, George traced the faulty line up a hill and across a field. Looking up along the course of the buried cable, be saw a well driller sitting in a chair in the middle of the field, watching his drill and listening to its rhythmic sound. As George approached, the man looked up and asked, "What are you doing here?"

"Looking for a break in the phone cable," George replied. "Did you water witch this well?"

"Yeah," said the driller. "What of it?"

"Well, you water witched the cable!" said George. And so he had.

The odds of drilling through a cable in the midst of a field by pure chance must be very low—but not zero. As they say, a stopped clock is right twice a day. Skeptics have tested dowsers scientifically, as in a 2004 test in Kassel, Germany, where plastic pipes were buried fifty centimeters under a flat field. The position of each pipe was marked on the surface of the ground, and all that was asked of the dowsers was that they say whether there was water running in the pipe. The results were no better than chance. Conclusion: German dowsers need to improve their skills. The data do not support any such conclusions about water witchers in Washington County.

The average farmer does not need such arcane skills as water witching or such specialized equipment as a drilling rig, but just about every farmer needs a tractor. Spring is the time to get the tractor out of the shed and use it for something other than plowing snow. Part of the allure of farming is the abundance of machines and gadgets that

The Skellie farm has land that runs all along the Battenkill River toward Center Falls. You can see the fields here in the early spring, and the furrows of corn that were harvested last fall. In late March, it is still too cold to begin planting anything this far north, and the fields are too wet to plow.

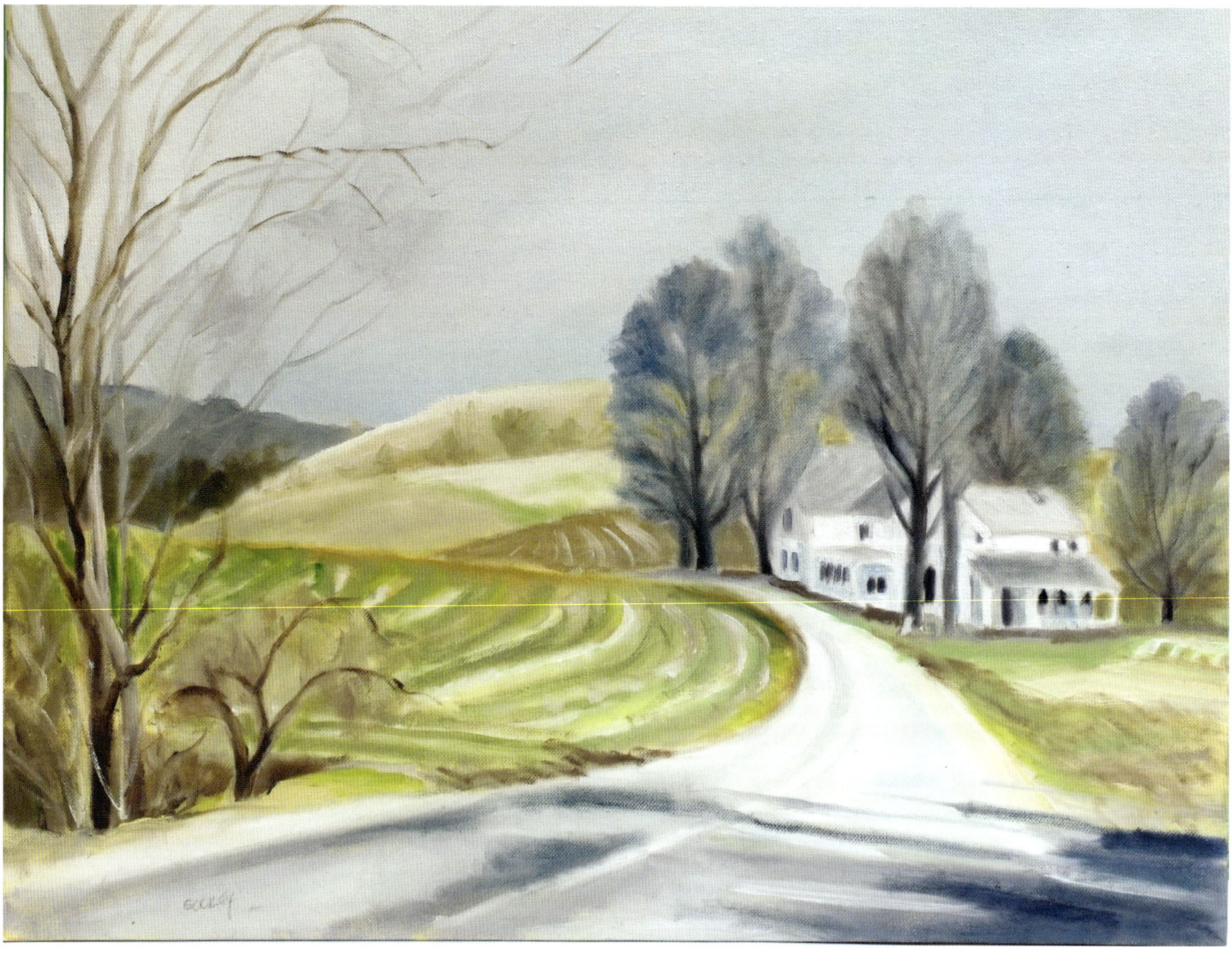

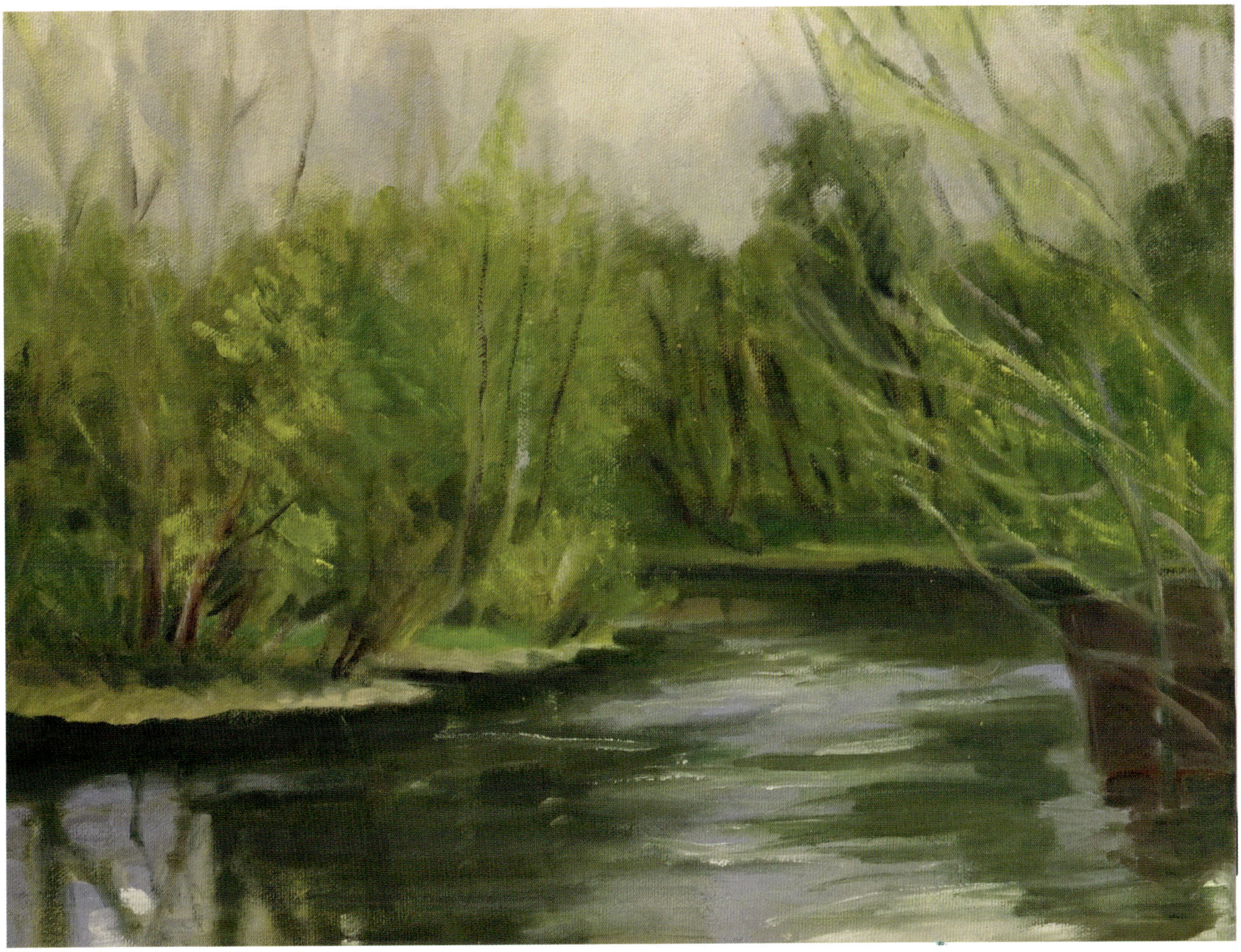

The Battenkill winds its way down from Vermont toward the Hudson, through farms that were built along the river valley, many over one hundred years ago. The farmland along the riverbank has a rich, dark soil, crumbly and fertile, perfect for many farmers who need to plant corn and other crops to feed their milk cows. This location is a favorite spot for the trout fishermen who go fly-fishing in the early spring.

make the farmer's work more efficient. The nineteenth century, that great age of mechanical inventions, saw a proliferation of hand-powered and horse-drawn machines for plowing, harrowing, planting, mowing, threshing, baling, and shucking. Many of the old machines have rotted away, abandoned in the weeds, but a fair number have survived, and collectors such as Rob Southerland at Gardenworks are putting them on display so that the next generation can see how things used to be done.

In the summer, the Washington County Fair will open its farm museum, with a good sampling of old tools. It will also hold two of my favorite events, the July show of antique tractors and the August tractor exhibit. Modern tractors, huge, powerful, and boxy, with enclosed, air-conditioned cabs and enormous tires, are impressive to see, but there is nothing like the old tractors. Even a railroad locomotive can't compare to the majesty of a steam tractor, with its smokestack, canopy, boiler, and steel wheels, prowling across a field. I admire these from a respectful distance; even today, they occasionally explode. Of course, if your time is up, a boiler explosion is a fine send-off.

The antique internal combustion tractors are the best of all. They are functional, affordable, and relatively easy to understand, and they don't contain computers of any kind. Let them sit around for a few months, and they still start right up. Dead battery? Just get out the

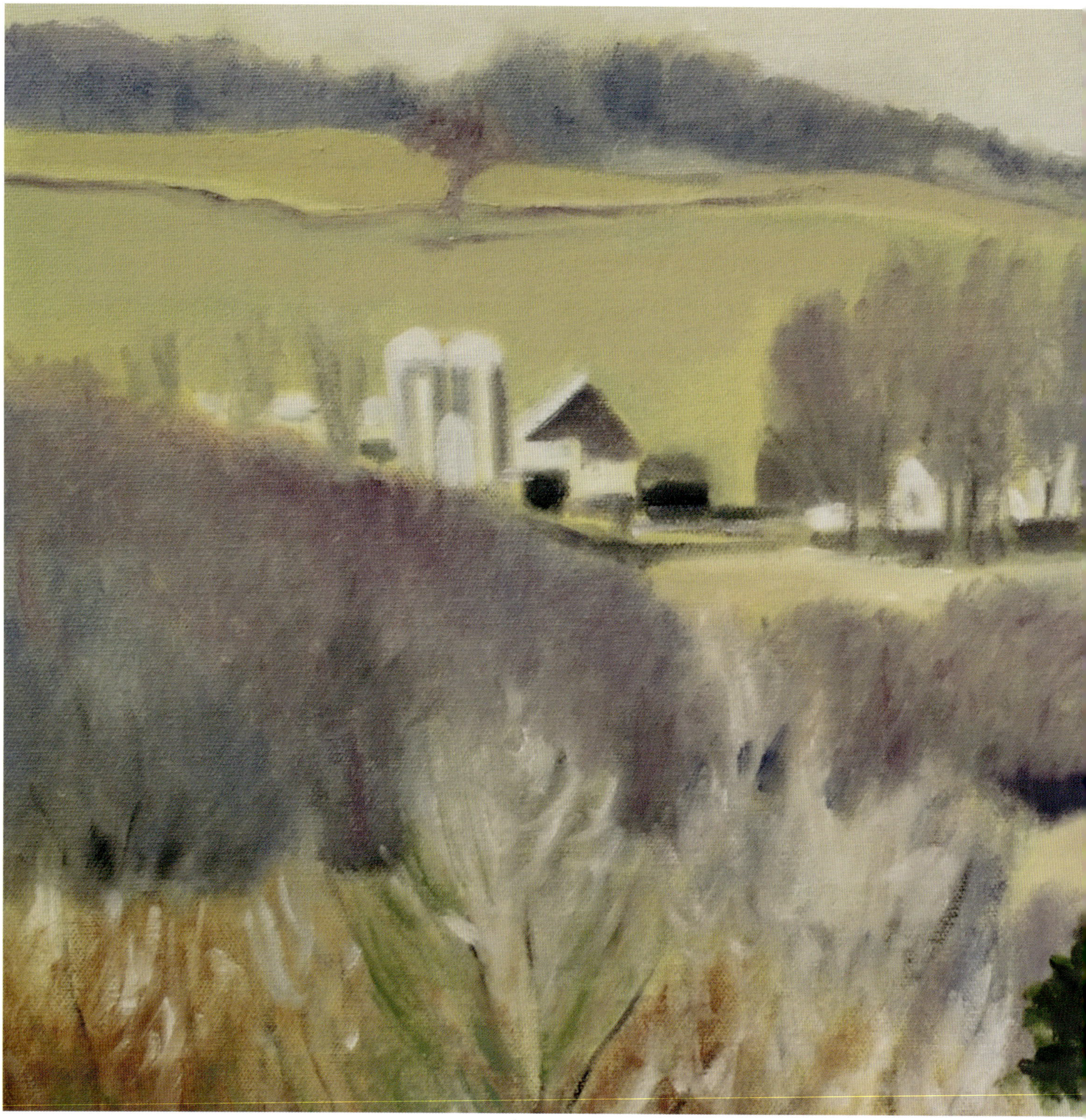

hand crank. A rogue regime gets the bomb and launches an electromagnetic pulse attack? Your car won't start, but the old John Deere will.

My own tractor in Talbot County, Maryland is a Farmall C. It isn't fixed up as nicely as the beauties at the fair—instead of red, it's Rust-Oleum primer brown—but it works well pulling a bushhog to keep the forest from taking over my field. With a little mowing, the meadow produces white daisies in May, yellow daisies in June and July, pink and purple asters in August and September, and white-tailed deer in November. Nature is bountiful.

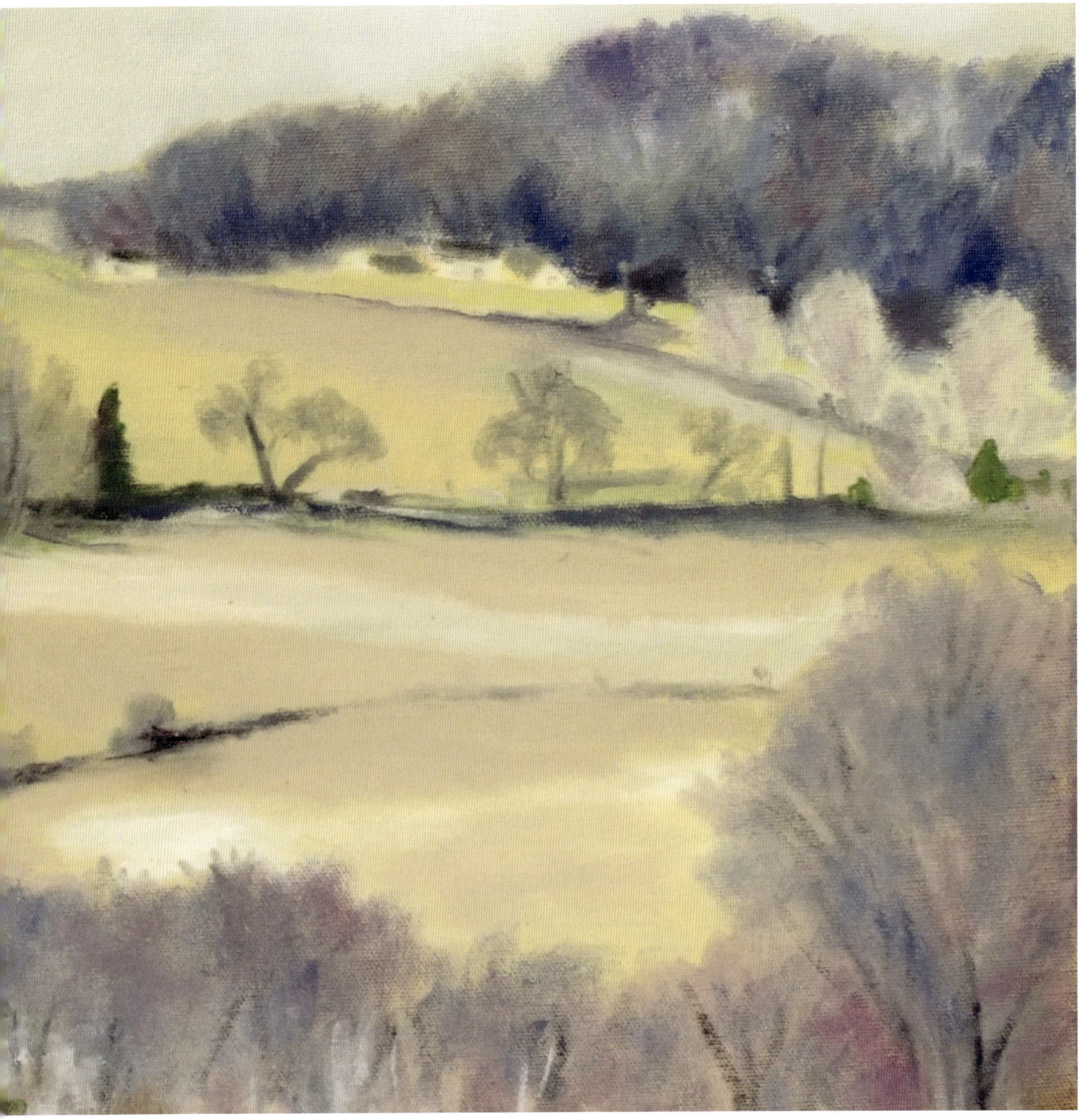

About halfway up on the climb to "Thunder Mountain," you can see farms in the town of Jackson on the other side of the Battenkill River. In early spring, the colors are mostly tan, white, and purple. It's cold too. The ground can remain frozen well into April, which is why it seems more like winter in early spring, even though the air temperature might be fifty degrees.

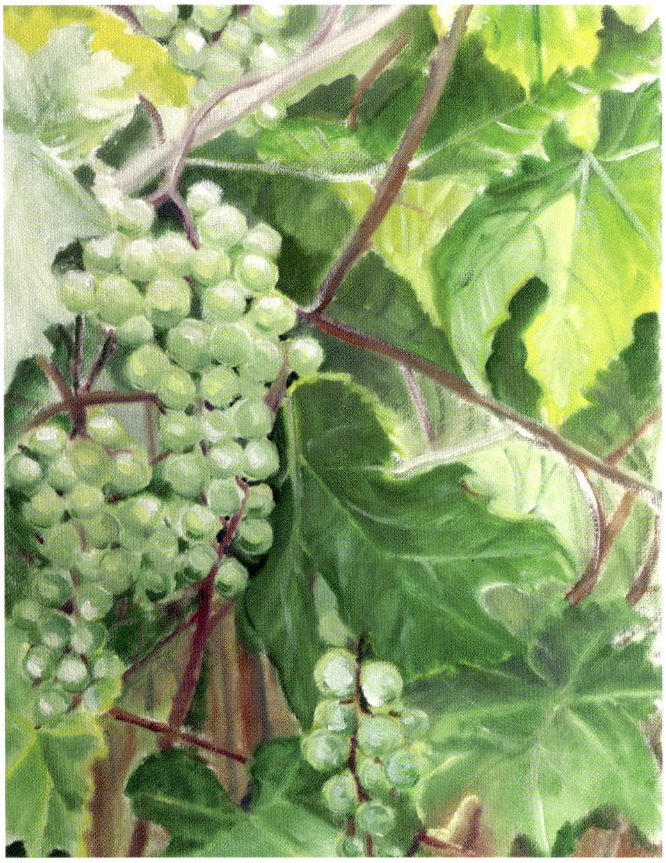

Amici Vineyard is located on the top of a hill overlooking Valley Falls on the south side. If you climb over a couple of the rolling fields on the property, you can also look into Vermont. Barton and I got a wonderful tour of the vineyard and saw the abundant grape crop. We were invited to sample some of last year's vintage too. Afterward, we had a bit of explaining to do to Alan Henderson, who found it amusing when we told him why we had been drinking so early in the morning.

SUMMER

The weather is warm, and the grass is growing. July is a good time to see how some farmers are getting back to raising chickens in a healthy, outdoor environment. That outdoor chicken farming seems like a novelty is a measure of how drastically food production has changed in the last half century.

After World War II, researchers at the University of Wisconsin reported that chickens fed on a diet of antimicrobials (in this case, sulfasuxidine and streptomycin) grew faster than those fed on a normal diet. It didn't take long for big agricultural businesses to use this information to increase the speed of production. Today somewhere between 50 and 80 percent of all the antimicrobials in the U.S. are not used to treat sick animals or humans but are added in subtherapeutic doses to animal feed. This dangerous practice goes on in spite of the fact that we have known for a long time that the surest way to breed drug-resistant bacteria is to expose them to low levels of antibiotics. Alexander Fleming, the discoverer of penicillin, warned of this risk in his Nobel Prize address in 1945.[8]

Subsequent discoveries have shown just how antibiotic resistance develops. It used to be thought that it was simply a matter of artificial selection of strains of bacteria with the right genes. Now we know that bacteria are trickier than that; they carry little packets of genes in "resistance cassettes," which may be exchanged between bacterial cells, even between germs of entirely different types. In this way, the gene for methicillin resistance could be lurking in some harmless member of the happy company of normal flora in the back of your throat, just waiting to help out the next *Staphylococcus aureus* that decides to attack you. A curable infection can then become fatal.

The June 2009 issue of *Johns Hopkins Magazine* describes some of the evidence that today's industrial farming practices are creating an enormous reservoir of resistant organisms. These organisms are carrying their little cassettes of resistant genes into every community in America, swimming through groundwater, riding on flies, and swirling in the dust that blows out of poultry trucks. There has to be a better way.

And there is. To find out about it, we drove up to see Don and Bridget Cavna, growers of pastured chickens, at Far Away Hill Farm in Greenwich. These chickens arrived as little fuzzy chicks, shipped from a supplier in Iowa. After a period of basking under heat lamps at ninety degrees, the chicks are weaned from the heat and released into movable cages, in which they graze on insects and plants in the pasture. Their manure improves the soil, and they benefit from the healthy open air and natural diet. They never need or receive antibiotics. And

8. Feiger, "Farmacology," 25–31.

best of all, when they are fully grown, eight to ten weeks after arriving at the farm, they taste better than any factory-farmed chicken.

Don and Bridget did not plan to become chicken farmers. Raised in Brooklyn, Don worked for many years on air-conditioning and refrigeration, never suspecting that the skills he acquired would serve him so well as a farmer. One year, during the winter slump in the air-conditioning business, he got a job offer from an old friend to work in ship brokerage, negotiating deals between ship owners and charterers. One high-paying, high-pressure job led to another, and after the opportunities in the shipping business became less attractive, he became a commodities broker. "I hated every minute of it," said Don.

All the time that he was working in the city, Don dreamed of having a retreat in the country. Since childhood, he had vacationed in the tranquil landscape of the Lake George region, and he hoped some day to recapture the sense of peace that he had felt there. Suddenly, his life changed when Lyme disease affected his heart. Twice, he came close to dying. A cardiac pacemaker gave him another chance, and he saw that it was time to stop putting off his dream of rural solitude.

The back road to Salem, New York, is charming and eventually leads into the town. In the distance you can see the mountains of Vermont. This scene is a favorite of my cousins Jack and Barbara Barber, taken from "Cemetery Corner." "We always wanted to buy that property and build a house there one day," Jack said. The original painting now hangs in their house, in its rightful place after all.

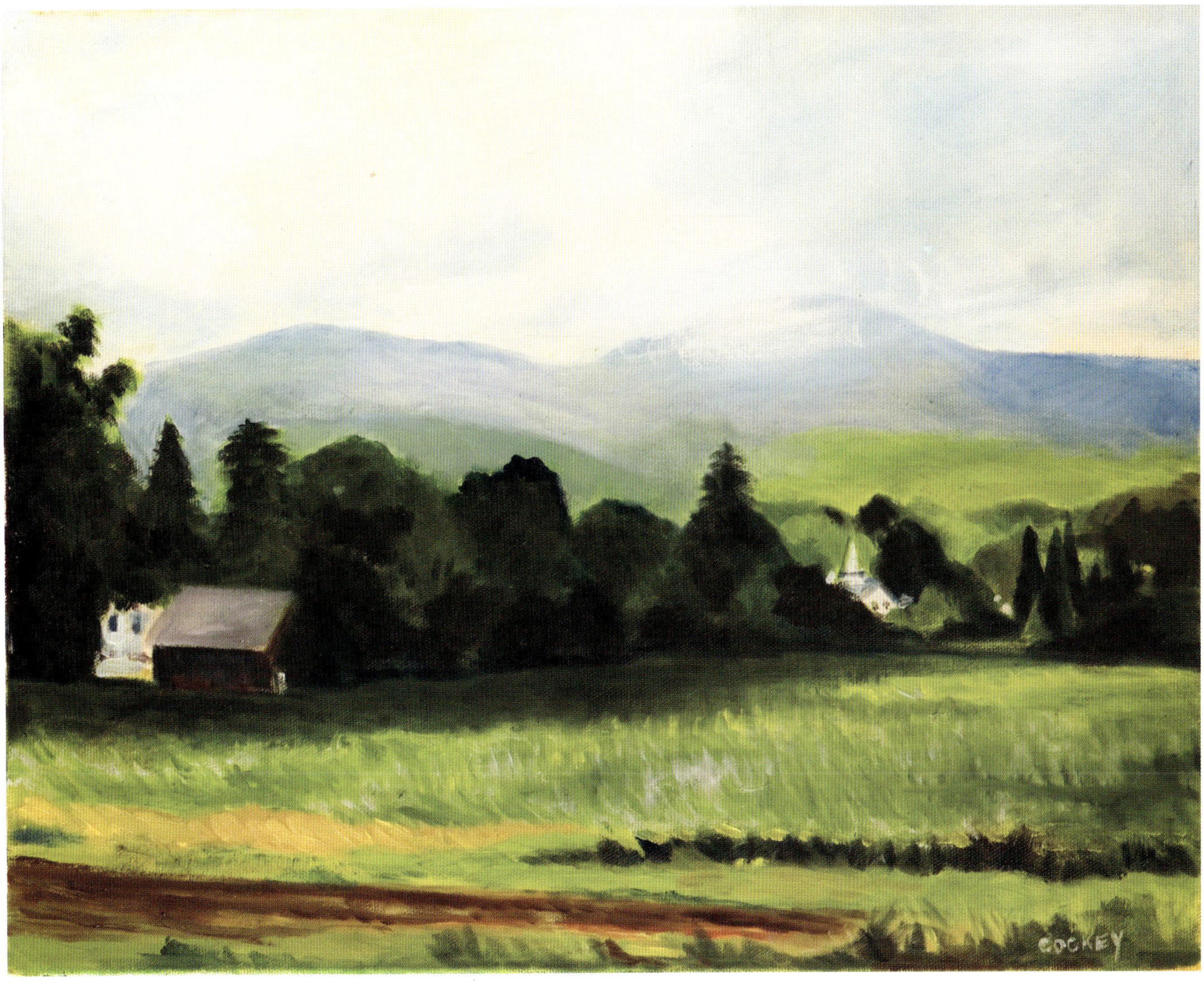

In 2001, the Friday after the World Trade Center went down, he and Bridget found their farm, an early nineteenth-century frame house on fifteen acres in the wooded hills, with a fine old barn and a pasture. There was not enough land for a herd of cattle, so chickens were a natural choice. Don had not had any prior experience raising poultry, but he figured it out. At the peak of his production, he was raising four hundred fifty meat birds a year. Now he has scaled back to about two hundred, selling to select customers, mostly friends, who place their orders in advance and pick up the freshly killed birds to stock their freezers. He also keeps thirty-five or forty laying hens.

The hardest part of the job is the killing and processing operation in the fall. Even with the help of Bridget, who brings her nursing training to bear on the task of maintaining a sterile environment for the meat, every year when he finishes the processing, Don says, "I've had it." When we asked what keeps him doing it, he said, "It's my friends. I like the interaction."

When we visited Don at his farm, I could see that this was a happy man. With a ready smile, a quick sense of humor, and a great New York accent, he immediately made us feel at home. While I admired the neat work he had done on an addition to his barn, Elizabeth talked with Bridget and watched the laying hens pecking around in the grass. As she watched, the hens stretched out on their breasts and spread their wings, forming a fanlike array of russet feathers on the brilliant green of the grass. Although Elizabeth grew up on

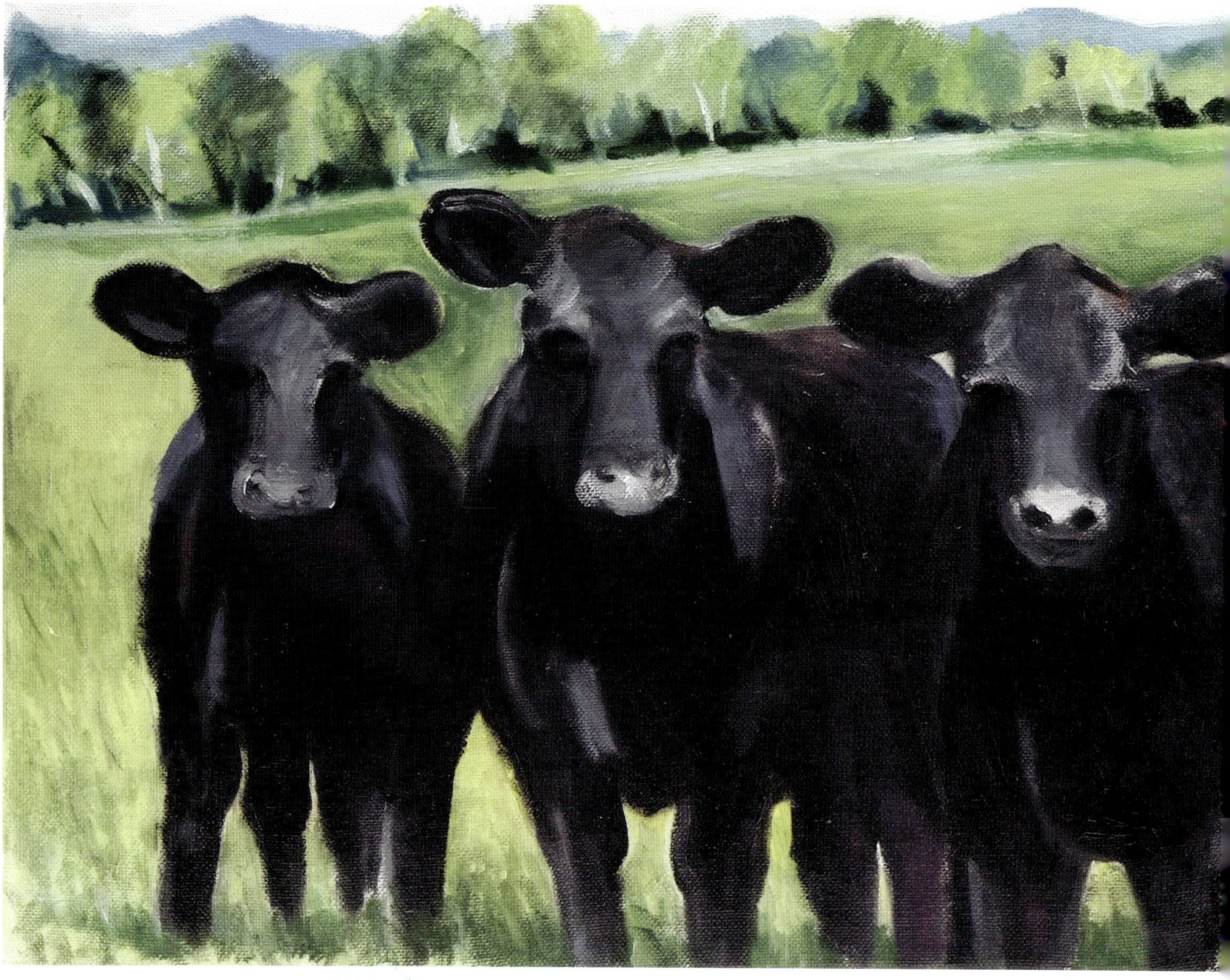

a farm, she had never seen this behavior in chickens. Bridget explained, "They're sunning themselves." These were truly relaxed fowl.

The horror stories of the beef industry go back even further than those of the factory chicken industry; Upton Sinclair wrote the first shocking exposé on the topic, *The Jungle*, in 1906. The latest concern in beef is the practice of fattening cattle with corn on feedlots before slaughter. Advertisements have so fixed the idea of corn-fed beef as a good thing in our minds that it comes almost as a revelation that cows are designed to eat grass. A concentrated diet of corn, often mixed with beef fat, feather meal, chicken litter, and other delicacies, tends to make cattle sick, mainly because it causes the contents of the rumen to become abnormally acidic. The sick cattle, crowded on feedlots, receive antibiotics. These practices fatten the cattle more quickly, increasing the marbling of fat in the steaks and lowering the price of beef at the cash register, but at what cost to our health?

To see how beef can be raised on a natural diet, we visited Kevin Jablonski and Karen Christensen

How do you photograph steers? Very slowly. What I did was hold my camera out in front of me, and the beef cows were interested enough to let me in close for this shot. The owners of Mack Brook Farm told me that I was the only person they knew who had done that without getting stampeded. Of course it does help if, like me, you've been raised on a farm.

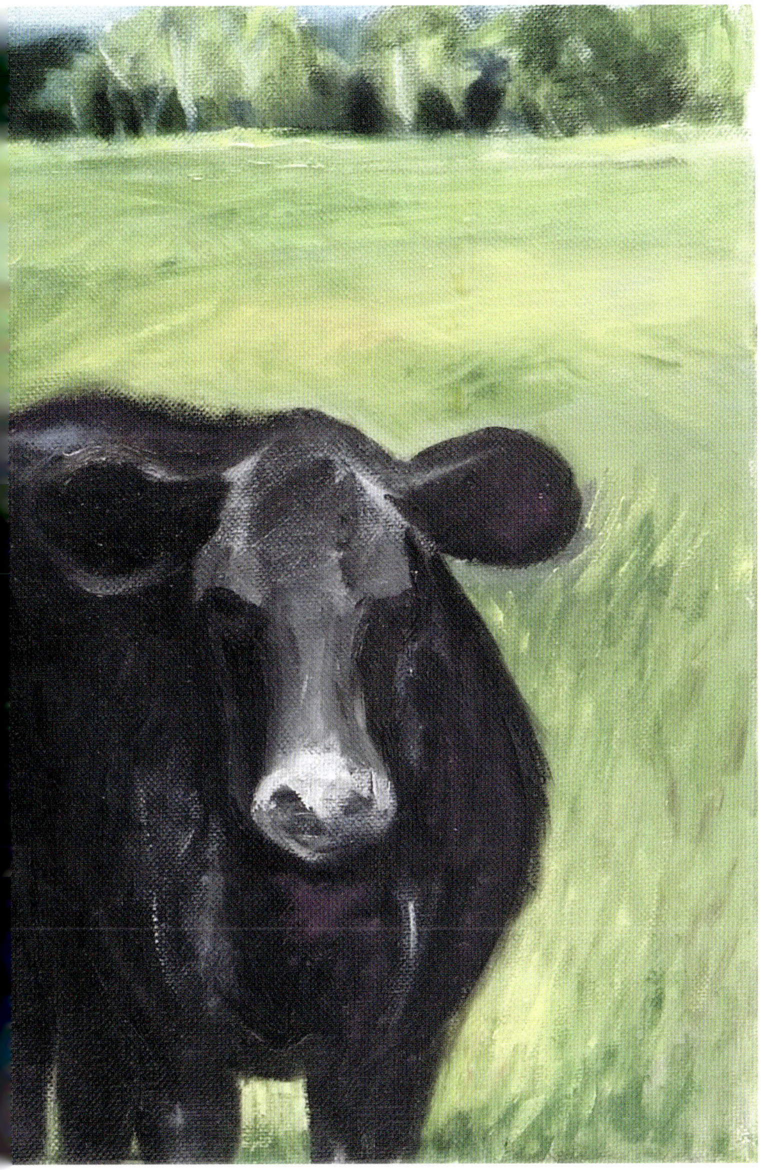

In Praise of 4-H

I pledge my head to clearer thinking, my heart to greater loyalty, my hands to larger service, and my health to better living for my club, my community, my country and the world.

—The 4-H Pledge

All over the country, 4-H has provided education, guidance, and constructive activities for young people for over a hundred years. It is a perennial survivor, outlasting many of the other voluntary organizations and benevolent societies that grew up in the same era. Devoted to the promotion of improved agricultural practices, its "learning through doing" approach encourages young people to be active participants in the modernization of farming.

About 1908, O.H. Benson, a school superintendent in Wright County, Ohio, devised the three-leaf clover emblem, the leaves representing "heart, head, and hands." Three years later, he added a fourth leaf to represent "hustle." Shortly thereafter, "hustle" was amended to "health." The change in the logo has not noticeably reduced the organization's drive and energy, as the bright young people and informative exhibits at every county fair attest.

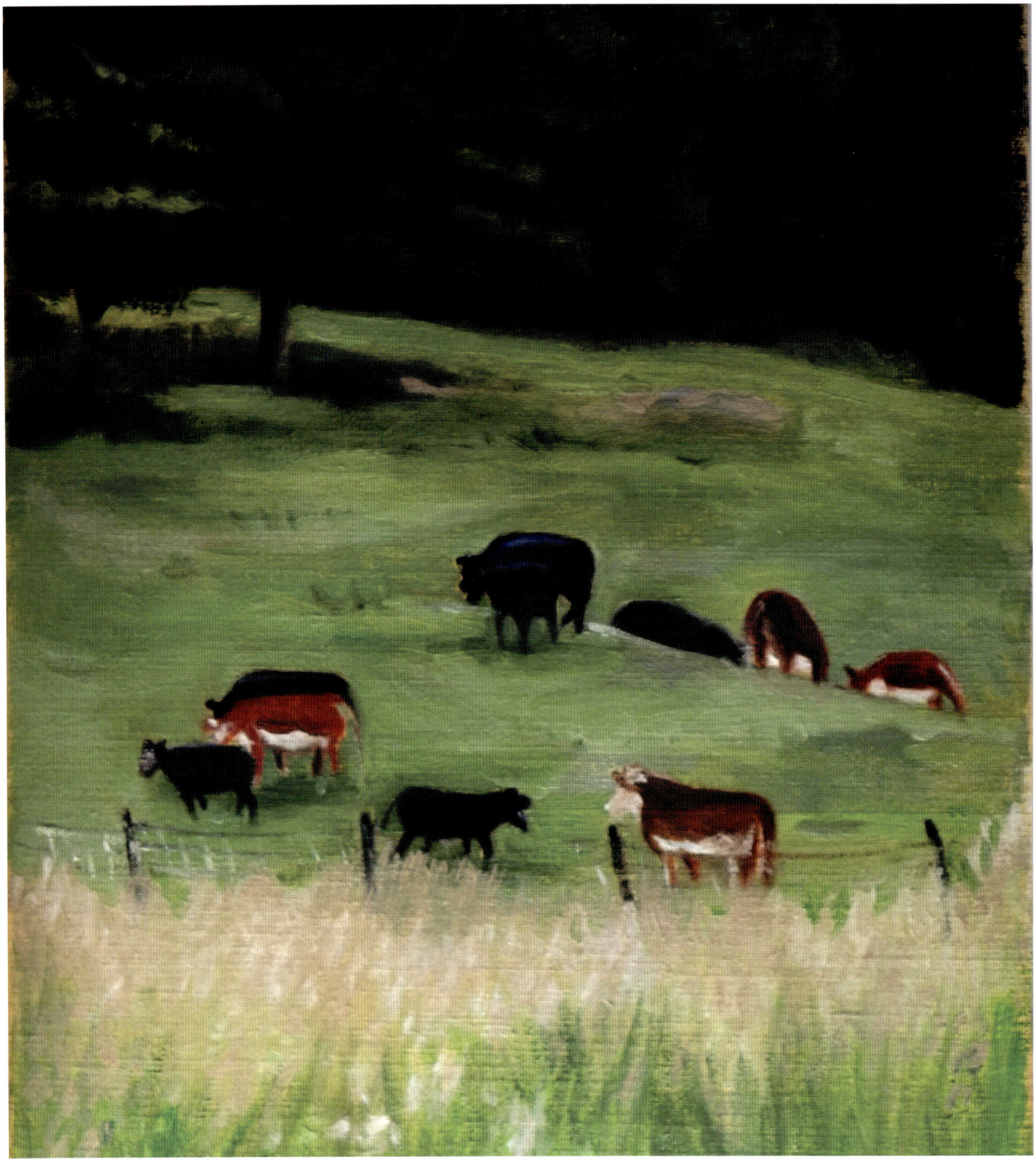

These cows are pastured on the west side of Ryan Road, which is one of our favorite drives through the countryside surrounding Greenwich. Everything you can see—the barns and farmhouses and the rolling hills—still looks the same as it did when I grew up here.

at Mack Brook Farm in Argyle. Kevin represents the third generation of a farming family. His father ran a local dairy farm until a serious automobile accident in 1966 forced him to sell his herd and rent his land to a neighbor. Kevin grew up with cows, raising young stock and working the farm in the summers when he was home from school. He always loved the beautiful

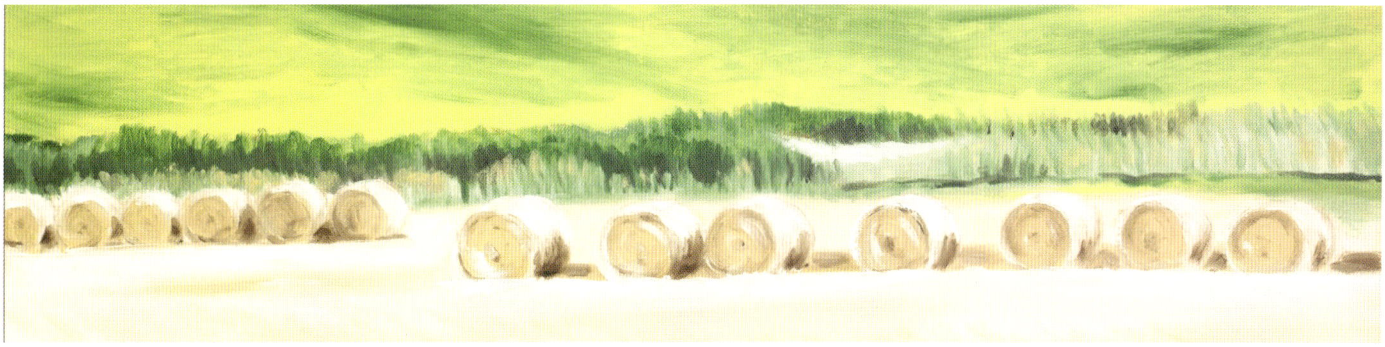

Above: I photographed these bales of hay on my way to Salem. On the return trip, they had already been gathered up and taken away by the farmer, and then it rained—a downpour. You have to move fast in these parts if you want to record local events. Last trip, tall corn was growing where the bales had been.

Below: Trout fishing on the Battenkill River is a popular sport from early spring through the summer months in upstate New York. In fact, the Battenkill River is favored by fly-fishermen who catch several varieties of trout there. This painting illustrates the unique relationship between the sportsman and the farmer. Many fishermen pull off the side of the road and hike down the bank to the river, where they are often seen throwing out a line. In most places, the farmland isn't off-limits or posted, and fishing seems to be an accepted practice on private property.

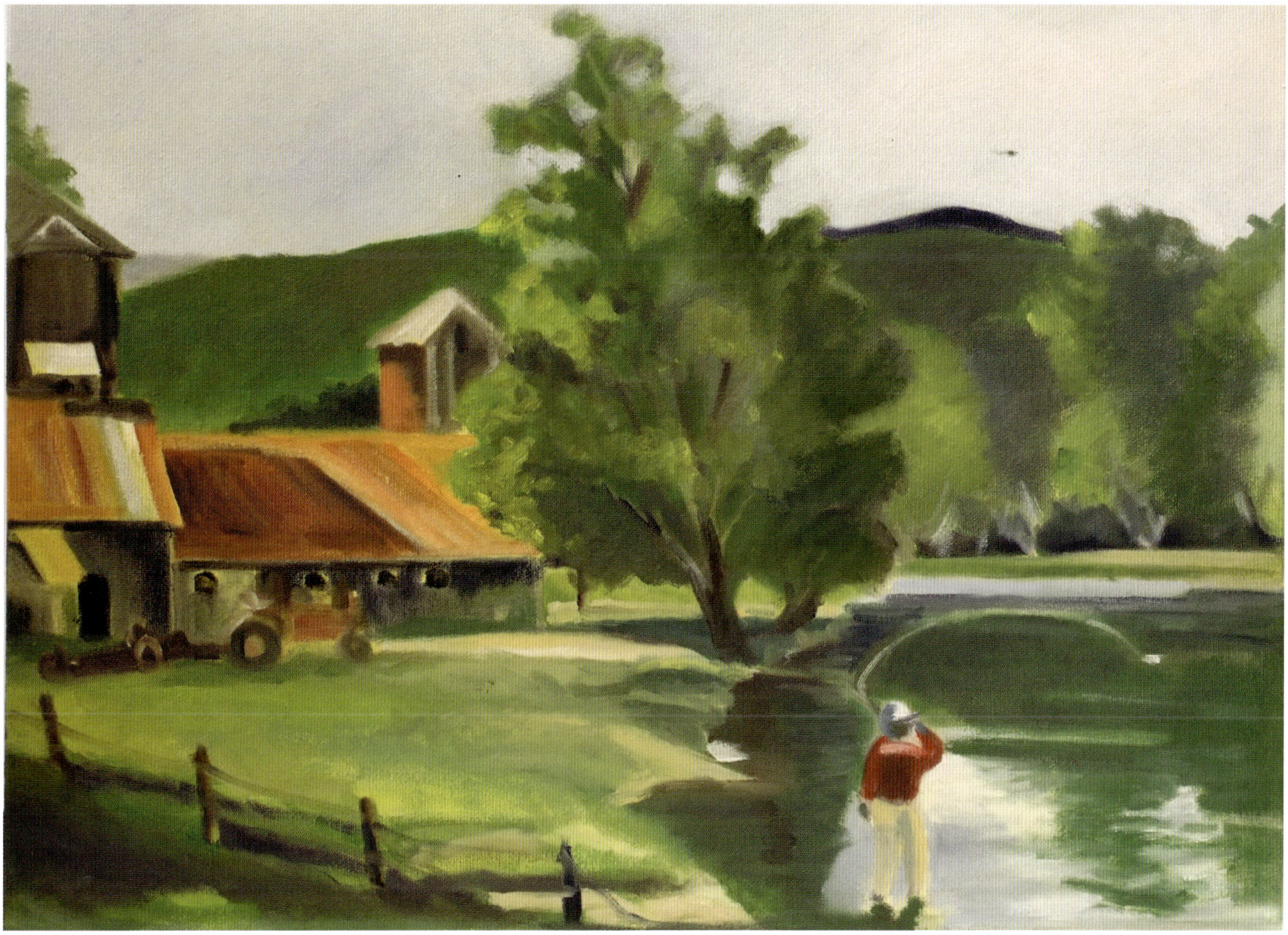

vistas from the hillside farm, across the Hudson River valley to the Adirondack Mountains in the west. At first, he was glad when his father sold the herd; they were a lot of work. But after a few years, he found that his feelings had changed. "I missed the cows," he said. "I wanted them back."

Planning and saving, he was eventually able to buy a good hay baler and wrapper and leave his regular job to devote himself to full-time farming. His herd now averages fifty head. He keeps the steers in fresh forage by rotating the pastures every day. By mowing and baling, he is able to put up feed for the winter. His Angus steers eat nothing but what grows in their pastures, and receive no hormones or chemical additives. The meat processor, Locust Grove, is just a few miles away, so the cattle do not become stressed or lose any body fat in transit. Everyone in these parts speaks highly of Locust Grove, but the facility is not certified "organic." The closest organic meat processor is so far away that the quality of the meat would suffer if the cows were transported there. Kevin believes in using "best practices." His operation is audited by Animal Welfare Approved and the American Grassfed Association. An organic

In upstate New York, the summer months are a bounty of color. These fields were planted with gladiolas and zinnias, creating a hedge of flowers alongside Route 22 on the way north. In this painting, the flowers nearly dwarf the house in the background.

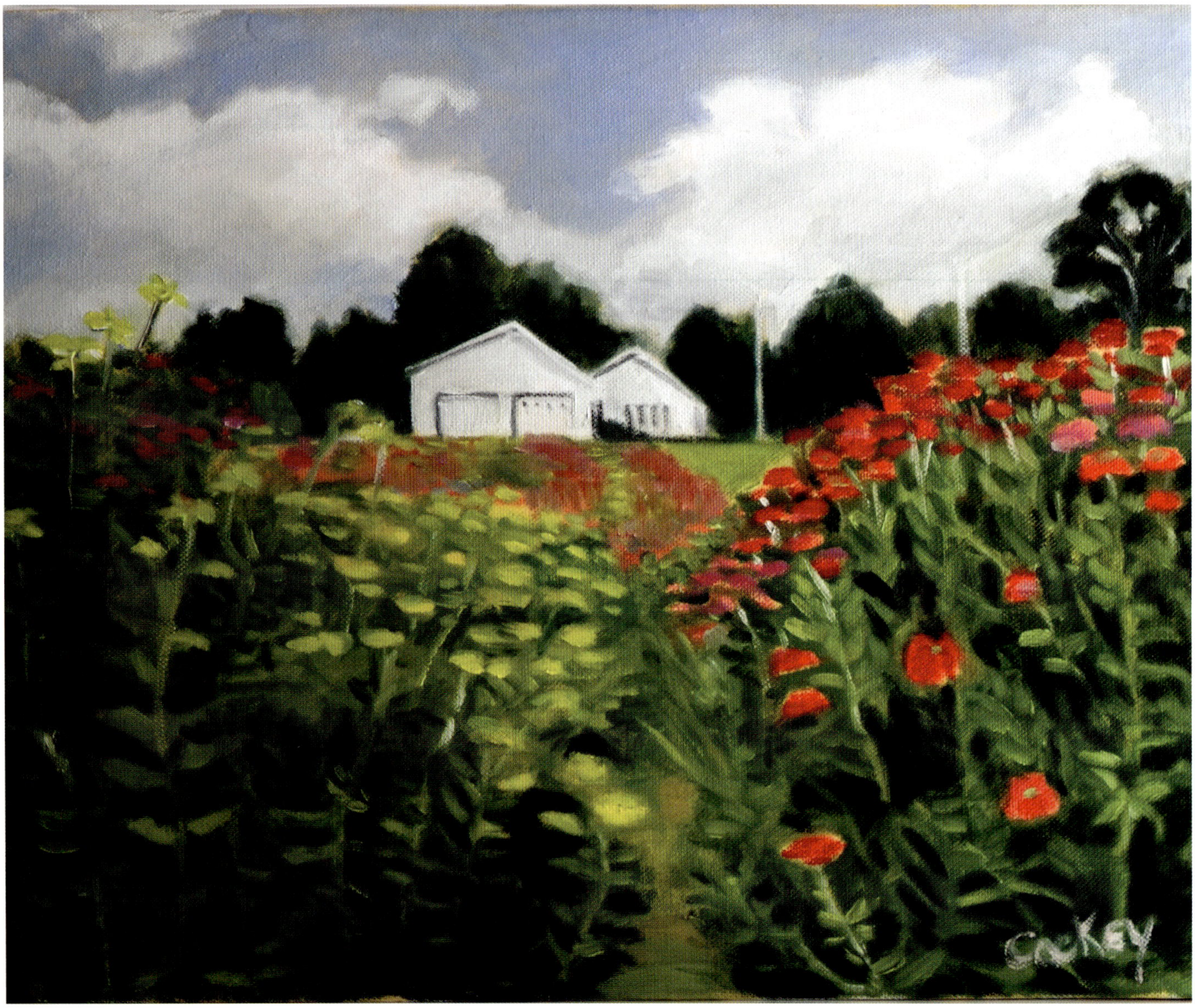

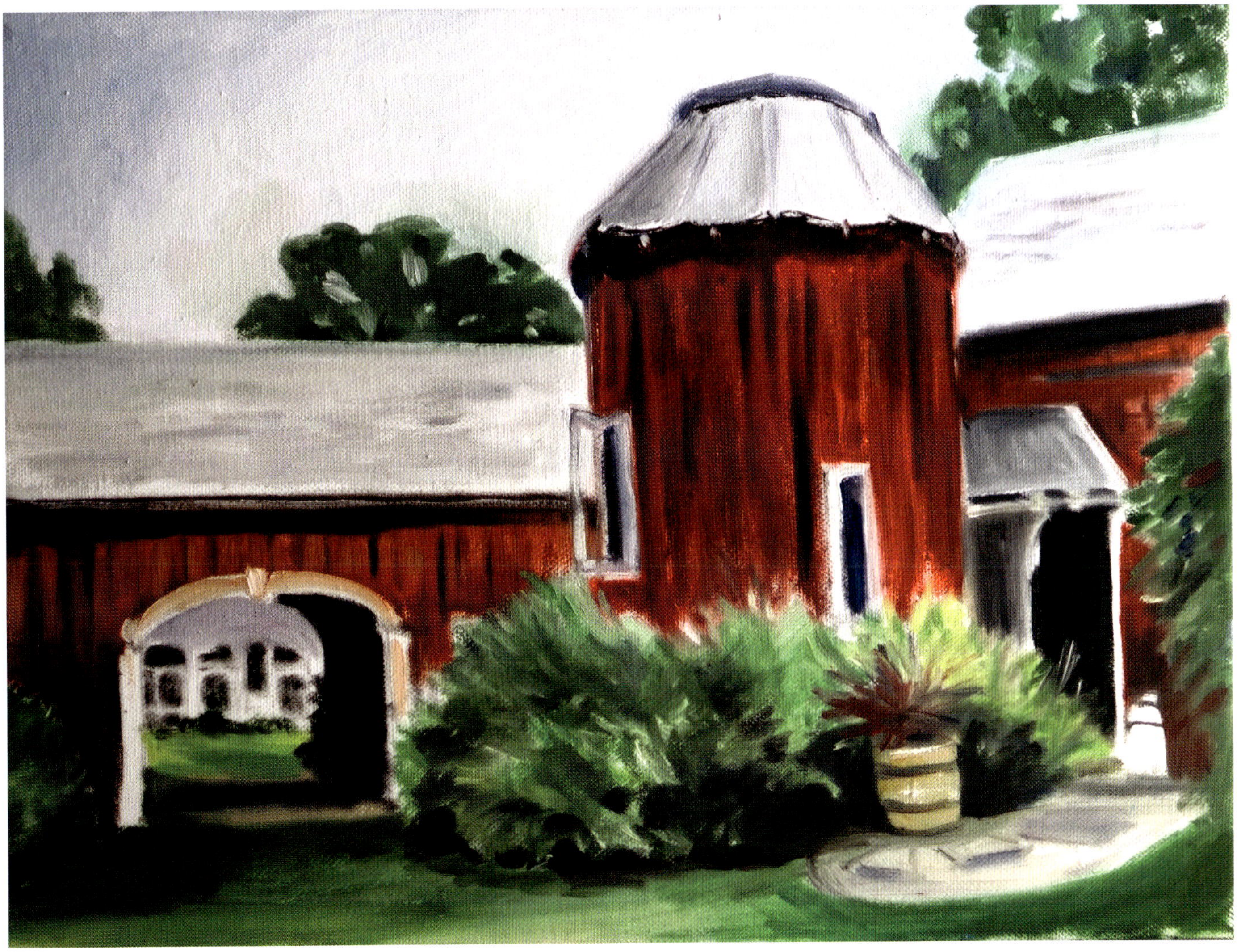

The red barns and silo are all part of Gallery 668's charming ambience. Originally built sometime around the late 1800s, they were restored to their original luster one hundred years later. Peeking out from behind the dogtrot of one barn is a lovely colonial home, which used to be the Battenkill Inn. Back in the 1950s, you could order a steak dinner for about five dollars. Oh, how I long for the good ol' days!

certification for this beef would be like a teachers-college certificate for Einstein.

Completing the romance on this farm is Karen. Her meeting with Kevin was as unlikely as any movie plot. He was installing milking parlors in Morrisville at the same time that she was dropping her daughter off at Colgate College. They both happened into the same bar, and they knew that they were meant for each other. After a three-and-a-half-year, long-distance friendship, Karen moved to the farm, bringing with her a passion for what she calls "using the land the way it's meant to be used."

"I have a real commitment to this kind of sustainable agriculture," said Karen. "We're going to change the world. This is the way to go!"

As we drove back to Greenwich, we admired the lushness of the corn; the rains had been good this year. Other than barns and grazing animals, there is surely no more iconic image of the farm than a wide, green field of corn, tassels waving in the summer breeze. And no wonder—corn is by far the largest food crop in the country. With an influential lobby in Congress, it is both beautiful and controversial.

Today, corn seems to be in everything. Aside from the obvious (corn chips, popcorn, breakfast cereals,

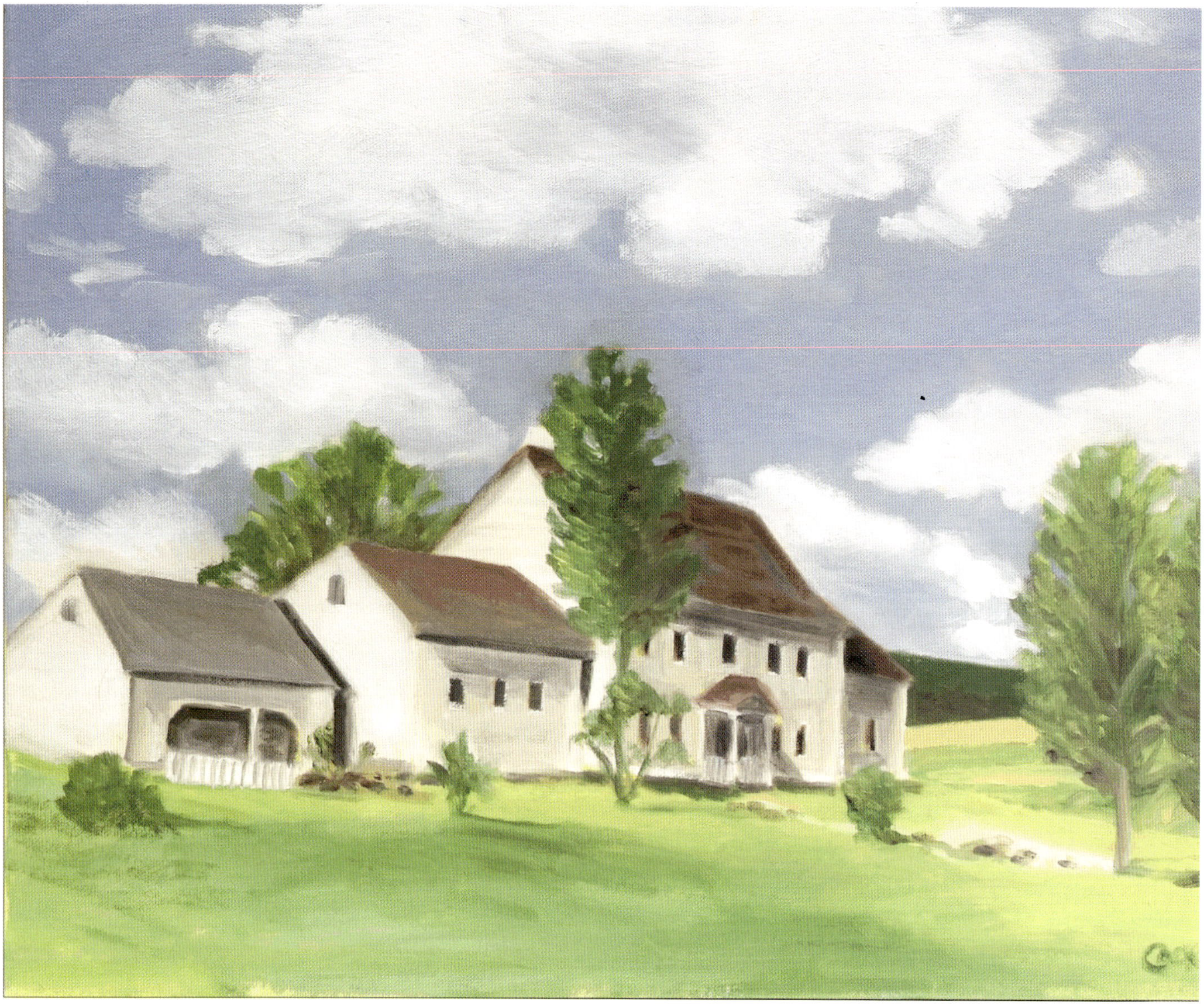

This lovely colonial home is being restored by the new owners, Mr. and Mrs. Jude Clary. It used to be a turkey farm owned and operated by the Petersons, and it was one of my favorite places to see when my parents drove to Cambridge, New York. Back then, the fields were covered with flocks of white turkeys. On this day, the clouds took up most of the scene, as often happens in summer.

corn-fed beef/milk and chicken/eggs), corn-derived products are in most processed foods. Corn syrup is a cheaper sweetener than cane sugar for soft drinks. Corn starch can be modified to form a moisture barrier between crust and tomato sauce in frozen pizzas, to stick the gypsum particles together in drywall panels, or to prevent ice crystals from spoiling the smoothness of ice cream. Paper, cosmetics, fuel additives—is there anything corn can't do? The list goes on and on, and thanks to research and development funded by the National Corn Growers Association (NCGA), the list keeps getting longer. More corn products mean more demand and higher prices, offsetting the effects of agriculture's perennial paradox: efficient production drives down prices. As the NCGA web site says, "NCGA corn research programs are focused on three main areas to fully leverage available funding to create processes that expand the potential of corn usage and develop profit opportunities for U.S. corn growers."[9]

Not surprisingly, the first point of "leverage" on their list is ethanol. You have to hand it to Big

9. National Corn Growers Association, "Research & Business Development."

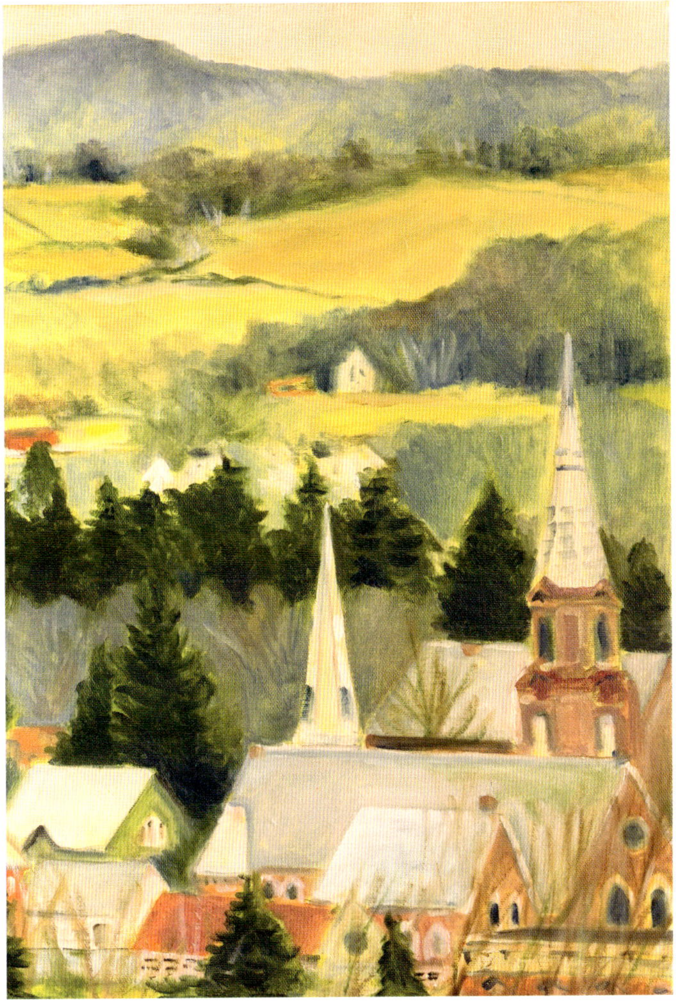

Left: This painting shows the unique relationship between Greenwich and the farming community that surrounds the town. Greenwich was originally a mill town, built along the river in the early 1800s while the outlying countryside was cleared and tilled for dairy farming. My family's farm was above the town, where we had a wonderful view of church spires, the school, and all the houses below us. I still feel very much at home and at peace when I admire the view from that vantage point on the hill.

Below: The Sheldons' roadside market is a wonderful place to find lots of their own homegrown corn and potatoes, but you can also snack on authentic tamales, organic cheese, and exotic treats, all made right in the Sheldons' kitchen. It is a true garden of earthly delights. There are other things too: antique books, china, and kitchen collectibles from the area.

Corn: persuading legislators to compel everyone to buy your product is a formidable marketing strategy. There's no choice; if you want gasoline, you have to buy the ethanol. At least your gas lines won't freeze. Now that the federal government regulates even such minute particulars of our lives as lightbulbs and showerheads, this strategy looks like a winner. One could view this development as a logical step in the evolution of corn. Here is a plant that is unable to reproduce without human assistance. It has given up its ability to survive in the wild, but it spreads by insidiously making itself indispensable to humans (have you ever wondered what makes peanut butter so smooth?) and now stretches in a vast monoculture across the Great Plains. Love it or hate it, no other plant has a higher stake in the continued survival of the human race.

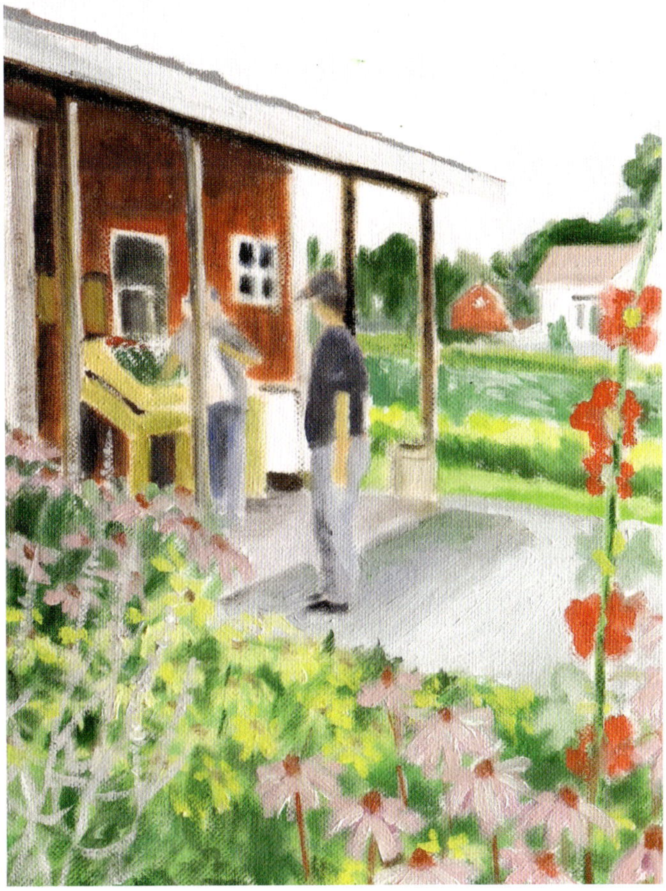

FALL

The season of plenty, of Thanksgiving, is also the time of the withering of the vines and the coming of the killing frost. It is the time to put away stores against the dire months of cold, when nothing grows but ice.

One of the best crops to put in the cellar for winter is the potato. When we visited the Sheldons' busy farm stand outside of Salem, we sampled some extraordinary food, including delicious hash brown potatoes cooked with terrine of duck, and we learned a few things about the potato farm.

Albert Sheldon comes from a long line of farmers. After landing in Boston, his ancestors made their way to Connecticut about 1640. They moved to Rupert, Vermont, in the mid-1700s and fought in the American Revolution. Around 1845, they migrated to Hebron and have farmed in this area ever since. Albert's father bought the current farm near Salem and often traveled to Castleton, Vermont to buy seed. There he met the daughter of the farmer who supplied the seed potatoes. She knew everything there was to know about the crop. Their marriage brought together many generations of New England farming expertise.

In this beautiful valley, Albert grew up with a love of the land. An avid outdoorsman, he hunts ducks and geese on his farm. He enjoys fishing. But most of all, he is an observer. A mink slinking among the cornrows, a killdeer feigning injury to lure intruders away from her nest on the ground, a flock of geese swinging in toward his decoys over a snowy field—these, for him, are the greatest joys of country life.

Pat Sheldon met Albert in 1990 through a mutual friend, an author, at a book signing. She remembers being impressed by his big hands. She immediately felt as if she had known him all her life. He reminded her of her grandfather in Germany; he had the same calm confidence and attention to the land. His barn, with its store of potatoes, called back to her memory the most pleasant places of her childhood.

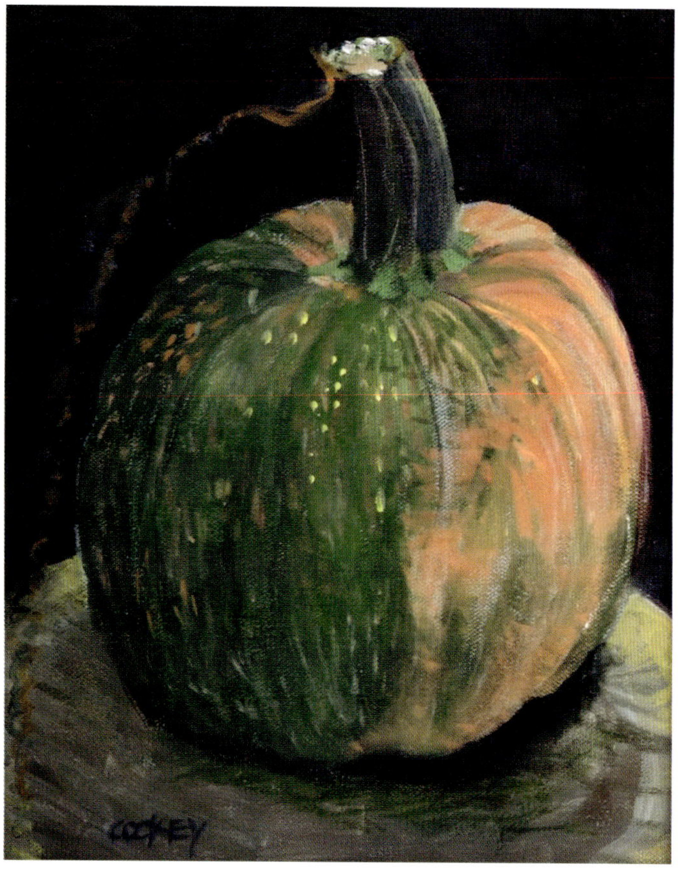

Volunteer pumpkins came up in Barton's and my garden. We watched them produce huge vines that nearly overran the tomatoes and other vegetables. All summer long, we waited. Surely there would be a huge crop of pumpkins! But no, there was only this one lonely little guy, half orange and half green. Since there was only one, I painted him.

Pat recalls her first time picking sweet corn on Albert's farm. They awoke and ate breakfast in the dark. Then Albert said, "It's time to go."

"How do you know?" Pat asked him.

"The first bird is singing," Albert replied.

They walked into the field as pale pink light suffused the eastern sky. Between two cornrows, Pat looked up to see a spiderweb, jeweled with dew. The first rays of sun slanted over the hills and shone through it. Such moments are her richest treasures.

And it is well that farm life provides such benefits, for the potato farmer has faced many difficulties in recent years. When President Clinton signed the North American Free Trade Agreement, American potato farmers watched

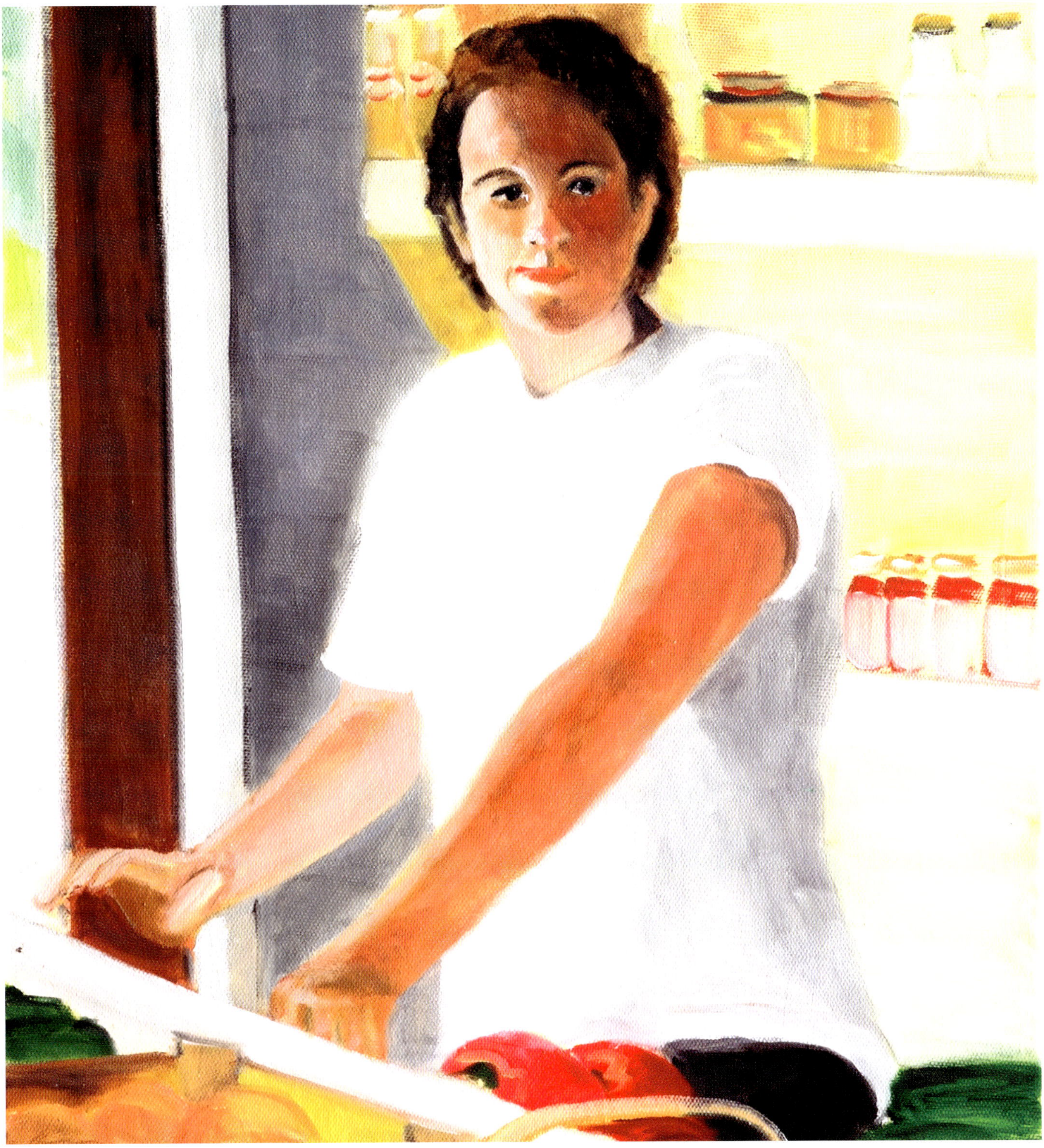

the price of potatoes fall as truckloads of cheaper spuds rumbled across the border from Canada. Public worries about fungicides, the Atkins diet and other food fads, and ever-increasing government regulations, fees, and inspection requirements have added to the trials of the potato farmer. "Not many people realize how close we are to subsistence farming," Albert told us.

On our way back from Williamstown, Massachusetts, Barton and I stopped in at the Moses Farm roadside stand, where we could see rows of corn, beans, and potatoes being harvested as we made our purchases. The girl behind the counter was very knowledgeable about the varieties of vegetables for sale and told us that they also made their own jellies and relishes.

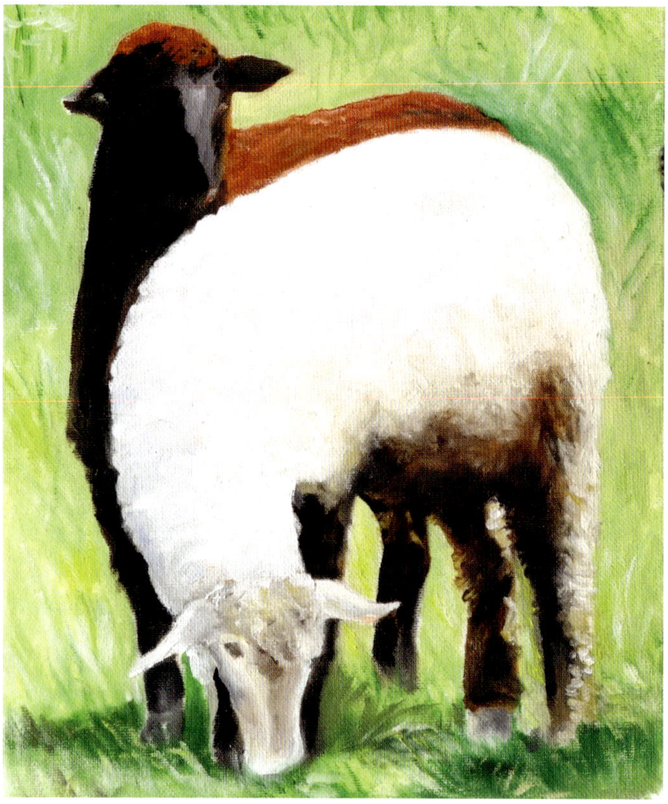

I was photographing Alan and Mary Lou Stern's farm toward the end of summer. The air was heavy with pollen, tiny flowers, and bees. Their pastured sheep came up to the fence and watched as we walked around the farm. These two seemed to be good friends, and, better yet, I got to pet them.

Fortunately, Albert and Pat have risen to the challenge. Diversifying their production to include a variety of vegetables, which they sell at a charming roadside stand along with old books, antique kitchen implements, and ephemera, they have found a way to supplement the farm's income. In addition, they are growing heirloom potatoes favored by urban connoisseurs. Even their plain red potatoes are above average. Pat attributes the superior flavor to the minerals in the soil here. A friend of ours made potato salad using the same recipe she had used for years. Her husband told her it was the best he'd ever had. "Well," she said, "that's because these are Sheldon potatoes."

My favorite potato story is about how the French scientist Antoine-Augustin Parmentier used reverse psychology to induce the French peasants to grow potatoes. Although the tubers were brought over from South America and introduced into Europe as early as 1536, most Europeans viewed them with suspicion. In 1748, the French parliament prohibited cultivation of potatoes, which the authorities believed caused leprosy. Meanwhile, Frederick the Great of Prussia was doing his best to use the coercive power of the state to force the Germans to eat potatoes. Thus, in an interesting variation of the statist principle that whatever is not mandatory is prohibited, what was mandatory in Germany was simultaneously prohibited in France.

When Parmentier was captured by the Prussians during the Seven Years' War, he had to subsist on a potato diet. Returning to France in 1763, he began to promote potatoes as a way to prevent the famines that attended the frequent failures of the wheat crop. None of his efforts were as effective as his ploy of posting guards around his potato patch at Sablons. The local people concluded that the plants must be valuable, and they stole them to plant in their own gardens. A similar ruse has been attributed to Frederick the Great, but Parmentier deserves more credit, as he has done the world a further service: his name remains as a clue to the presence of potatoes in many puzzling items on French menus.[10]

In America, there was no need for tricks or regulations to establish potato culture. The practical Irish had planted potatoes as a field crop since about 1663, and Scotch-Irish settlers introduced potatoes to Londonderry, New Hampshire, in 1719. The plants appeared in Connecticut by 1720 and in Rhode Island by 1735. Rhode Island had enough to export by 1745.[11] But it wasn't until 1853, at the restaurant of the elegant Moon Lake Lodge in Saratoga Springs, that one of the greatest potato products emerged when chef George Crum learned that a diner had rejected his French fries as too thick. He prepared a thinner batch, only to be rejected again! Determined to teach this finicky guest a lesson, Crum sliced the potatoes paper-thin and fried them so crisp that no fork could spear them. To his

10. Histoire en Ligne, "Antoine-Augustin Parmentier."
11. Prince Edward Island Potato Museum, "The Potato Industry."

surprise, the guest was delighted, and the potato chip was born.[12]

Driving west from Salem into the Black River Valley, we stopped to see Meg McEachron Southerland at Gardenworks on County Route 30. A beautiful farm with tall white barns when viewed from afar, Gardenworks revealed even more of its charms on a closer look. Sheep were grazing near a little stream that ran past the barnyard and the farm store. Chickens clucked excitedly when we arrived.

The field behind my parents' old house affords a view of the entire valley below, including the towns of Greenwich, Jackson, and Easton. Willard Mountain lies in the far background. I grew up on "the hill" with a half-mile driveway that we walked every day to catch the school bus. It was always lovely there any time of year, but especially in the fall.

12. The Great Idea Finder, "Potato Chip History."

Meg never expected to return to the family farm. When she went off to college, her father, with her brother's help, was managing the place just fine, as his father had done before him. He liked keeping dairy cows and chickens, but watching the price of eggs go down made him decide to renovate the fine old chicken barns for cows. They're a smart family, those McEachrons, as anyone in these parts will tell you, with farming in their blood; various relatives own several farms in the vicinity.

Meg had different ideas, but her farming genes were strong, and she earned a master's degree in horticulture. In the course of her studies, she met Rob Southerland, who was a student of landscape architecture at Longwood Gardens at the same time she was there. This year, they celebrated thirty-three years of marriage. During those intervening years, they lived in western Massachusetts, Philadelphia, and Kentucky, but the longer they stayed away, the more Meg missed the hills and fields of home.

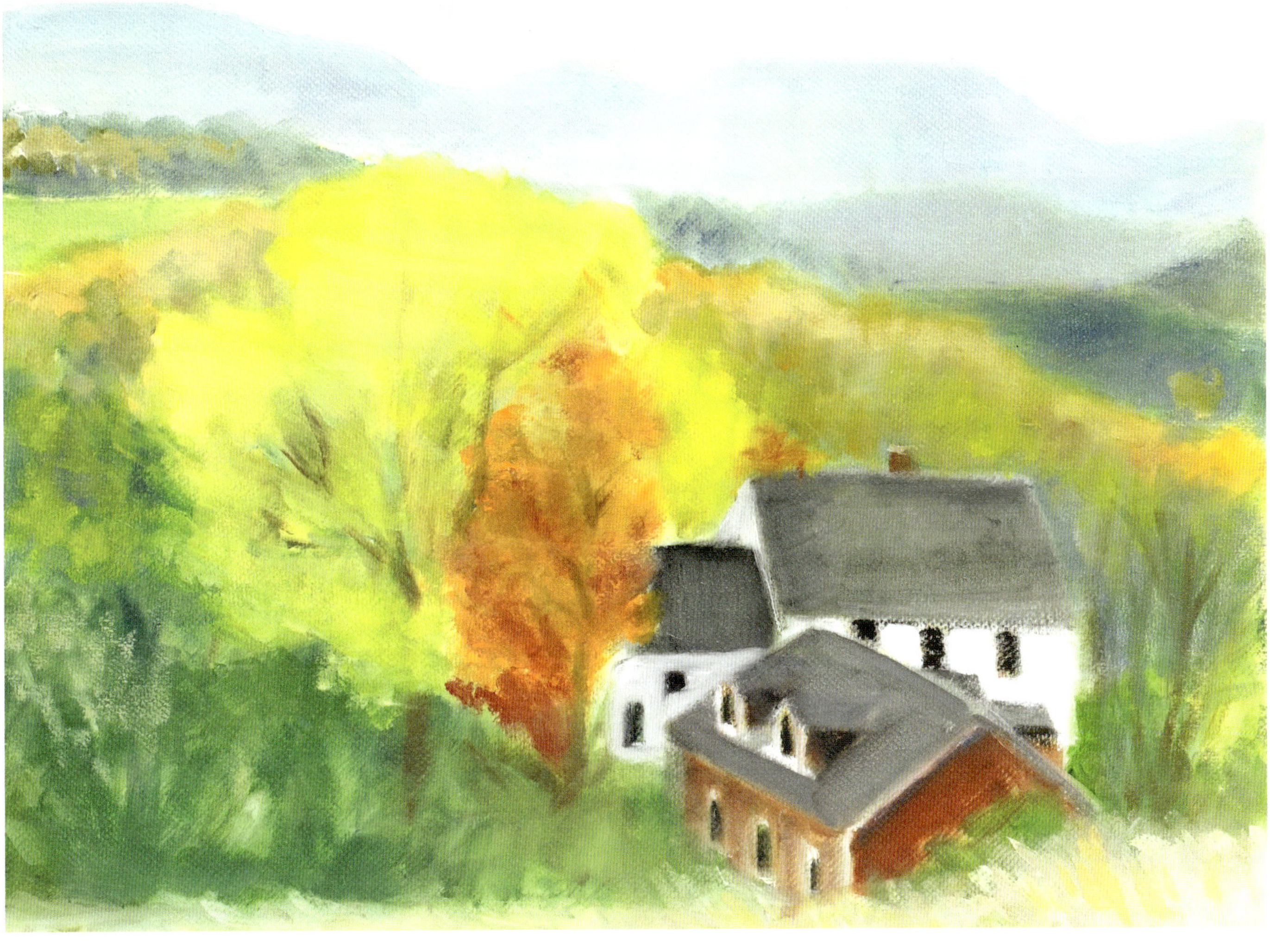

Every time Meg and Rob came back to visit the farm, it was harder to leave.

Now they faced a dilemma. They were secure in Kentucky—Rob was a tenured professor—but they missed the Northeast. It had those rolling hills, and you didn't have to run the air conditioner in the summer! Their opportunity came when Rob visited a landscape architecture firm in Saratoga. They offered him a job, and in 1990, Meg came home again.

While working as an extension agent in Kentucky, Meg had seen a farm that had changed over from growing and drying tobacco to doing the same thing with flowers. In the early '90s, dry arrangements were in fashion, and Meg began cultivating flowers, drying them on the second story of one of the barns, and selling them in a store that she and Rob created on the first floor of the same grand old post-and-beam building.

The store has grown and is now a four-season destination with a vast array of healthy annuals from the greenhouse in April and May, an abundance of pick-your-own berries in the summer, squash and pumpkins

Route 40 will eventually take you to the towns of Schaghticoke and Stillwater, and the city of Troy. It is a winding road and one of the most pleasant drives heading south that I know of—much nicer than the throughway! All along the way, there are farms with barns and cows such as this one.

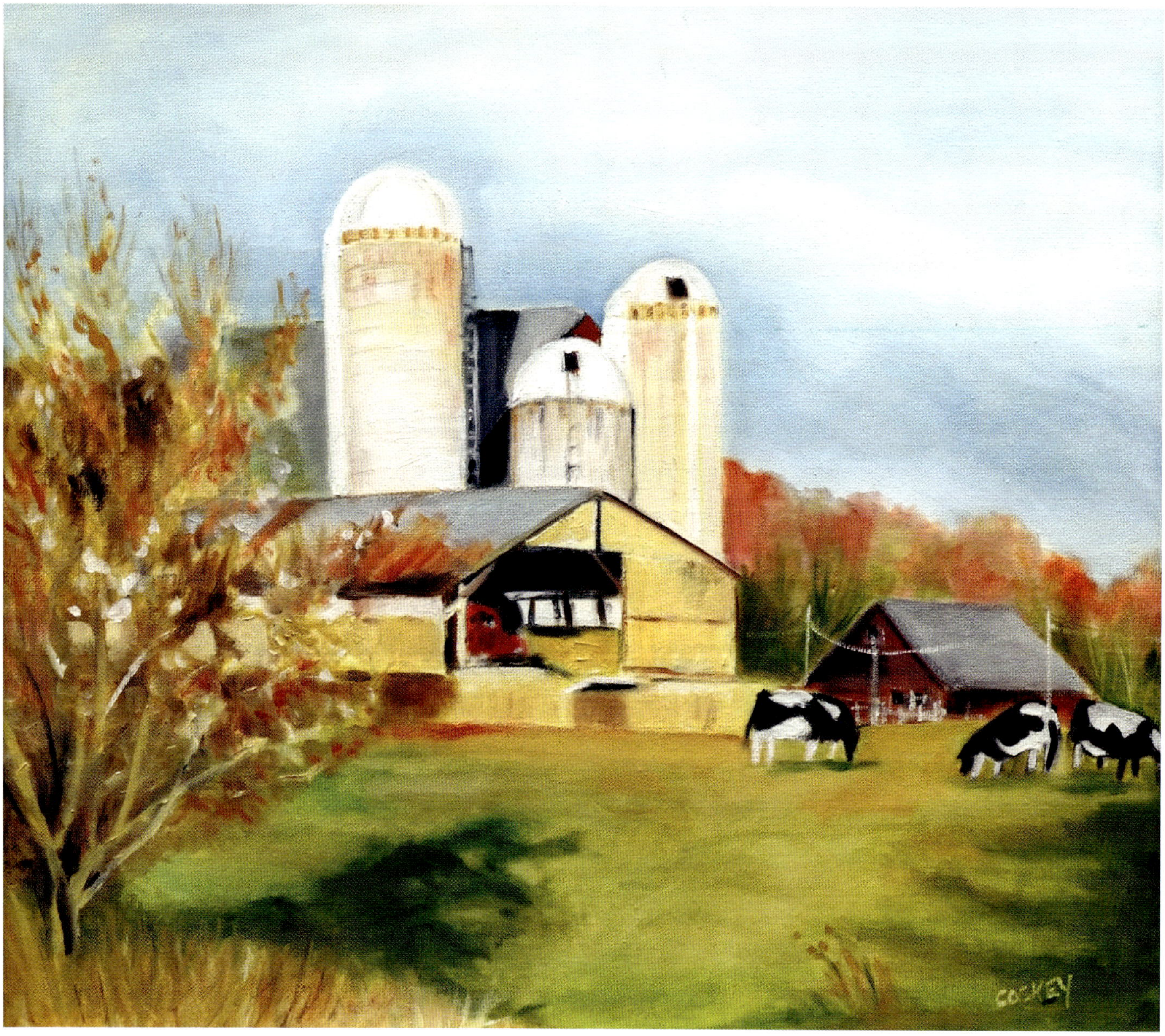

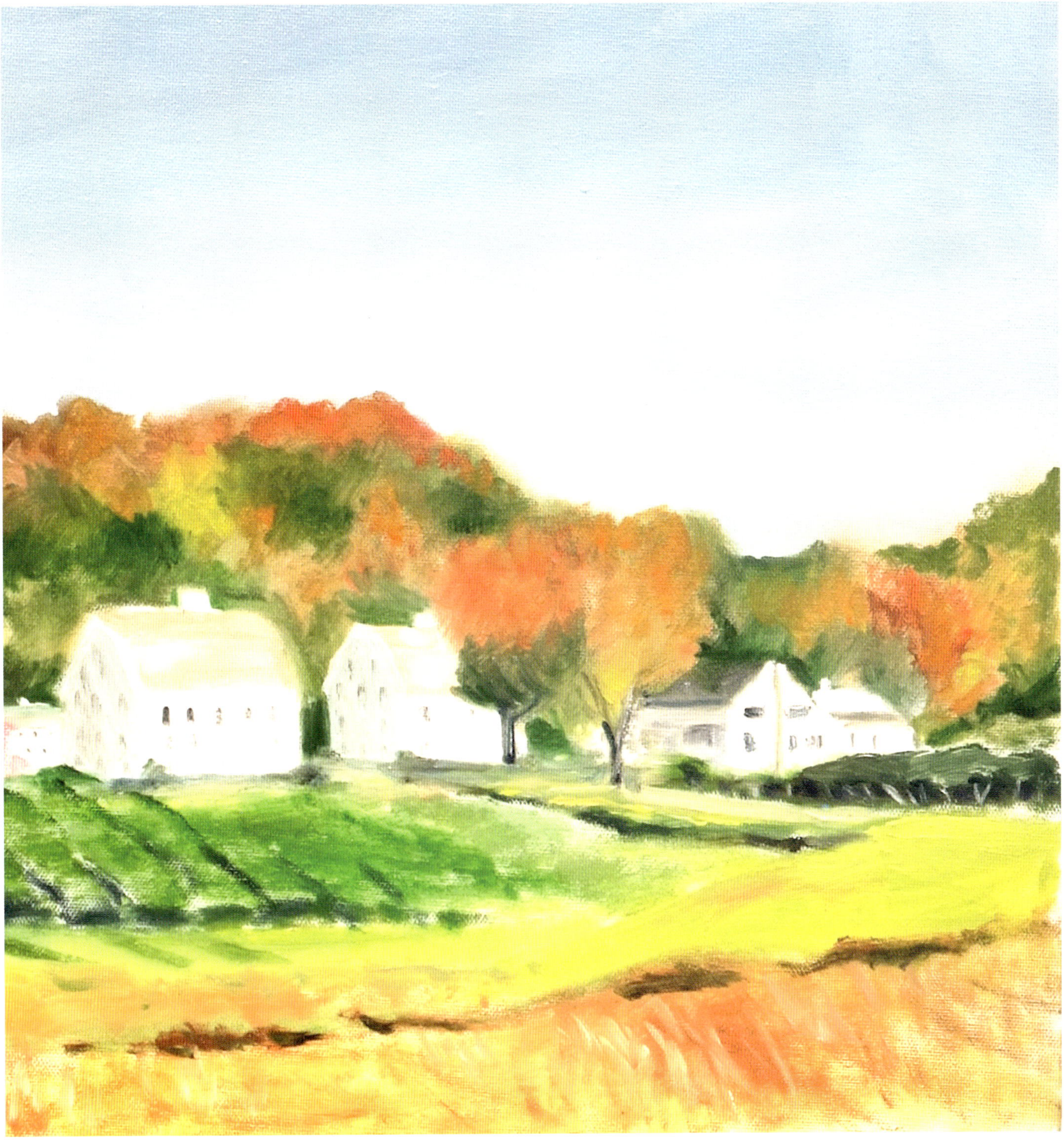

in the fall, and a Christmas wonderland of handmade wreaths and decorations in December. There are pies, preserves, and homemade breads. The refrigerated cases are stocked with local meats, cheeses, and eggs. In the store and on the second floor, Rob has assembled a fascinating collection of old farm equipment.

"We have the best organic vegetables," said Meg. "We don't grow them, but we know who does. I help them,

It's a small world, especially in a small town where everybody knows everybody else. Meg Southerland, owner and manager of Gardenworks, went to college with my first cousin, Carol Smith. They shared a common love for botany and gardening. Today, Gardenworks is a destination farm where you can pick your own berries, create a natural wreath from herbs, or browse the farm stand for local produce, pies, cheese, and potted plants.

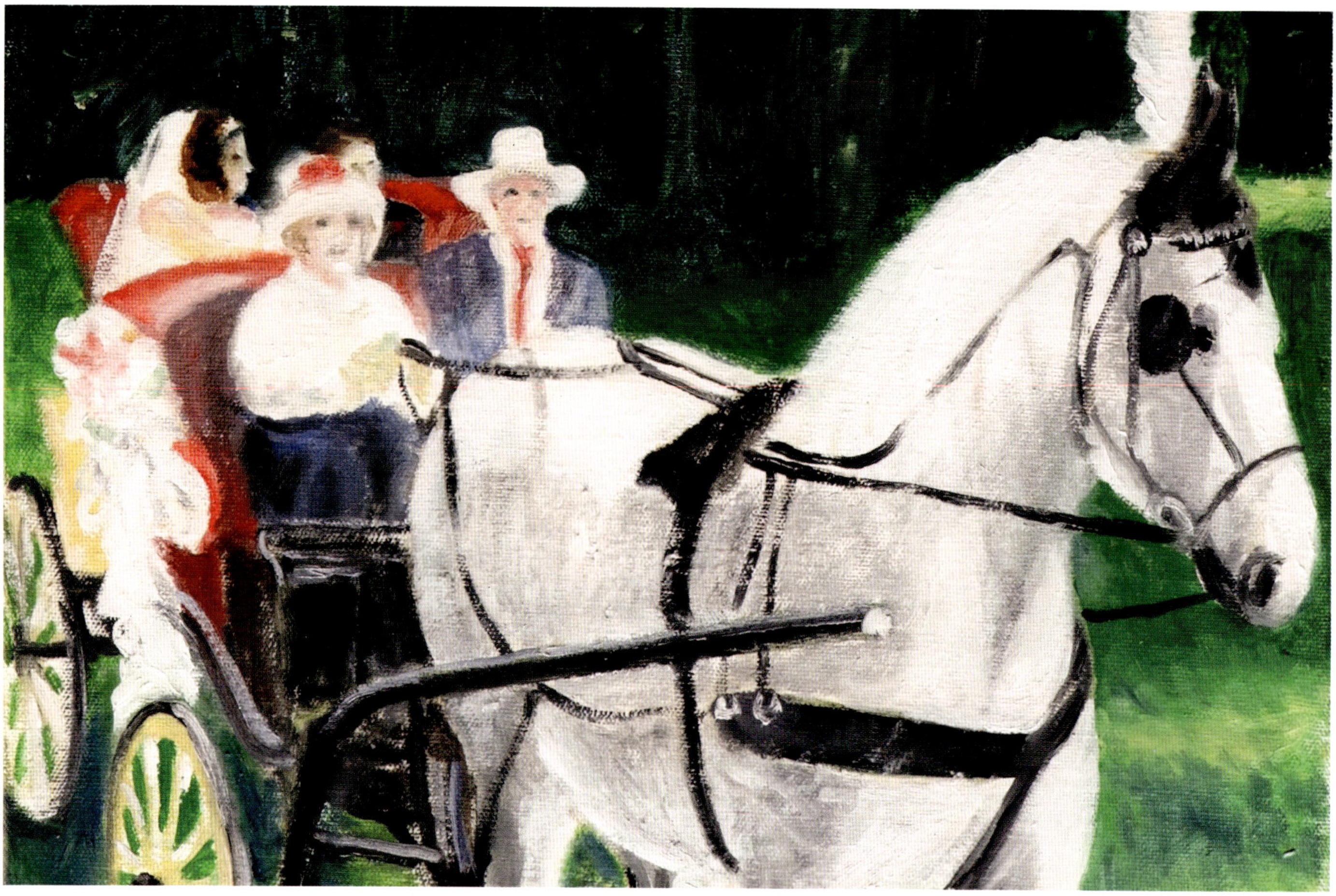

Emily Van Driel's horse-drawn carriage is well known in these parts. This day, they were carrying a bride and groom back to a wonderful reception out in the country. I know this because I passed the gala event several times when I was driving around lost. I spotted the horse's plume first from quite a distance away and just couldn't resist taking the picture.

and they help me to make this a more interesting place." Meg speaks with deep feeling of the importance of a community that works together, each member helping and sustaining the others. Here is her home, and here are the people she loves. The death of her brother has made this place and all the lives around her even more precious. The farm has helped to fill the emptiness created by his death. "My heart is here," she said. "Rob's job allowed me to listen to my heart. If you're willing to work hard, and you love what you do, it will all come to you."

Meg is not the only McEachron to be drawn back to the land. Her cousin Seth, after studying business and accounting at Skidmore College, has returned to his father's dairy farm. His knowledge of business principles has helped the family to develop its own bottling operation. Everybody told us that we had to visit the Battenkill Valley Creamery, saying, "Try their ice cream!" and, "They have the best chocolate milk in the universe!" So, of course, we went, and we found out that everybody was right.

The farm uses no synthetic growth hormones and raises its own feed. The decision to do their own bottling required research and careful planning, but the popularity of the creamery has exceeded all expectations. "People want to get back to natural and know where their food is coming from," explains Tracy McEachron, Seth's mother. The store has expanded to accommodate the demand.

There is something fundamentally right about eating ice cream within sight of the herd of contented, cud-chewing beasts from which it came. Of course, not being a dairyman myself, I couldn't tell which ones were the chocolate-mint-ice-cream cows.

The same terrain that lends itself so well to grazing cattle also offers ideal conditions for the cultivation of grapes. Grape harvest is a time for celebration, as the

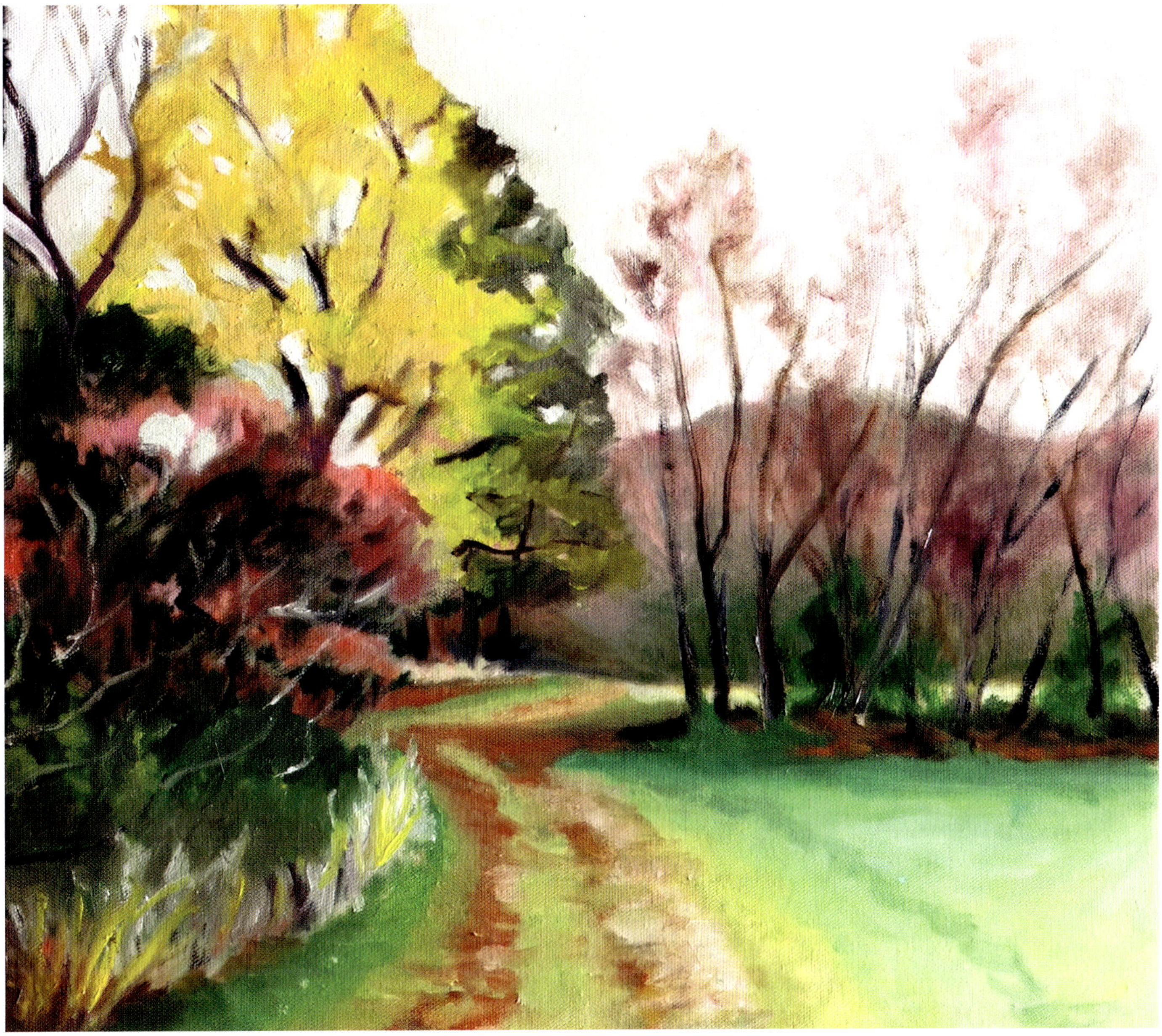

This lane is truly my favorite place to walk. There are still the remains of stone walls, fallen down now, all over my family's property. Two or three generations ago, my grandfathers would plow the fields and then drag them with a stone boat, digging round rocks out of the earth. They also found a wide variety of Native American artifacts: stone necklaces, broken pipes, and arrowheads dating back several thousand years.

vines yield up their bounty and the vintner improves upon it. Human craft and the secret workings of yeast transform raw fruit into a higher and more mysterious substance. When we raise a glass of wine, is it custom alone or something much deeper that prompts us to salute our friends and give thanks for the communion of kindred souls?

The soil and climate of New York are well suited to the cultivation of grapes. Hilly country is ideal, and there are hills aplenty. Atop one such hill, on a dirt road named after a Colonel Burch, is Amici Vineyard, the creation of gourmet chef Joe Messina. Joe is the epitome of the Italian gentleman-farmer-vintner. Tall, handsome, gracious, and ebullient, he gave us a tour of his vineyard and cellar. Rows of vigorous vines stretched along the hilltop, laden with clusters of succulent fruit. Joe plants Frontenac, Marquette, and Kay Gray grapes, all hardy varieties resistant to the cold winters of upstate New York. Frontenac grapes ripen to a deep blue-black and

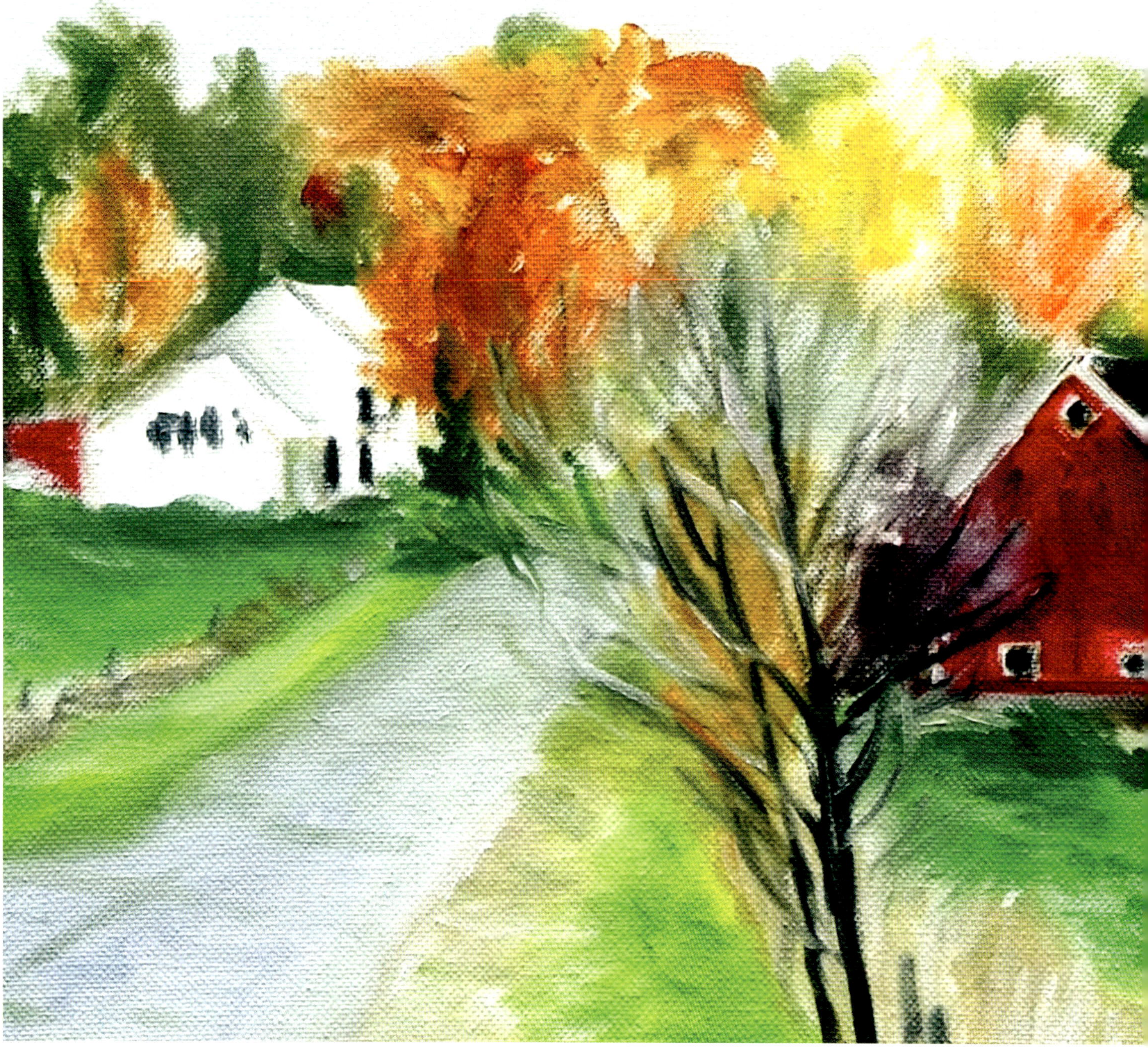

Take the North Road from Greenwich to Cossayuna Lake sometime. It is a wonderful up-and-over drive with lots of cultivated fields on both sides. The cows pastured there are well known for their high milk production. Every once in a while, you can still see ancient apple orchards that were planted over one hundred years ago still bearing fruit.

can be made into rosé, red, or port wine. Marquette, said Joe, is "the cold-hardy answer to Cabernet Sauvignon." Kay Gray is a white grape, well suited to eating or winemaking. Gazing out over his vineyard, Joe told us, "My lifelong work was to buy this bit of property, and the rest of the work is to steward it. And it will sustain me in my retirement."

Inspired by Walter Taylor, who advocated honesty in winemaking, Joe adheres to the "Old-World style," using no chemical additives. "My mission," said Joe, "is clean wine."

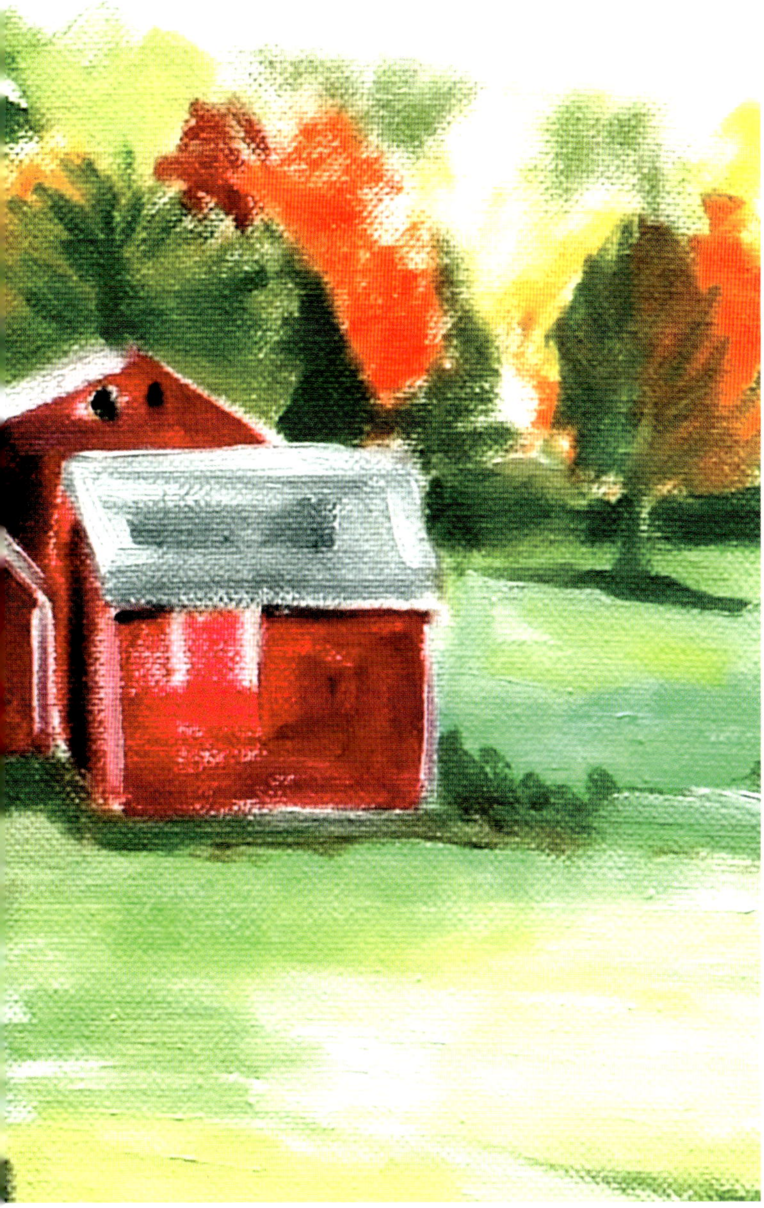

Amici Vineyard turned out three thousand bottles last year and expects to produce forty-five hundred this year. Joe says that at six thousand bottles a year, the operation will be "sustainable." A winemaker has many challenges to face. In addition to marauding deer and fungal blights, there are the taxes: fifty-four cents a gallon federal and thirty cents a gallon state, plus a 7 percent sales tax. Then there are the regulations, which change according to the size of the winery, and the record-keeping requirements. The vintner must record all the grapes he grows and all the sugar he buys. But the beauty of the land soothes the soul. "I live in heaven," said Joe.

In the immediate vicinity are two other established vineyards and two more in process. Joe does not see them as competitors. "My vision," he said, "is to pull all our local wineries together and work as a cooperative, help one another, market together. My experience is in restaurant sales. If I'm driving my truck and ten cases of wine to Saratoga, I want to drive their wines too and sell everyone's wine."

George Gilder wrote recently about the golden rule of capitalism: "The good fortune of others is also one's own." This truth is self-evident to a generous spirit. As Elizabeth and I basked in Joe's hospitality, tasting his excellent Riesling, Cabernet Franc, and rosé wines with bread and cheese in his kitchen, we gave thanks for good drink, but most especially for good friends. Amici Vineyard is well named.

As autumn draws to a close, we celebrate old friends and new, and eagerly anticipate the next cycle of snow, seedtime, growing, and harvest, all set against the backdrop of forested hills and rich river valleys. The farmland calls us for the next round of the dance. One striking image of the farm country that does not appear in this book is nevertheless indelibly etched in my memory. We couldn't record it before it ceased to be; it is to the artist as the "big one that got away" is to the fisherman. A few years ago, there was a place near Greenwich where black-and-white goats wandered among young birch trees. This idyllic study in black and white seemed like a metaphor for something—but if it was fraught with meaning, we never figured out what meaning it was fraught with. There was a strange piquancy about the scene, a hushed foreboding. Then, with the grim inevitability of Greek tragedy, the voracious goats chewed through the papery bark of all the birches, girdling the entire grove at goat-mouth level. Next, of course, the trees died, and the goats went away, we knew not whither. Make of this tale what you will. For us, it was "the landscape that got away."

CONCLUSION

Many of us are drawn to the beauty of the rural countryside. Those who choose to be more than mere spectators and actually work a farm express their affinity for farm living as a fact of biology. "It's in the blood" is the expression we heard the most often. Some used a more modern formulation: "It's in the genes." Regardless of where the trait is actually carried, all agree that it is felt in the heart.

Current statistics suggest that the trait is dying out. Only about 960,000 people in the whole country claim farming as their principal occupation. A similar number of farmers claim some other principal employment. Farmland is shrinking as more is sold each year for development. The trend continues, even as the state buys up development rights in order to keep farms in production.

Yet as the seasons bring their cycle of renewal, the land beckons to new generations of farmers. Even as some move off the land to seek other jobs, others with a strong commitment to good agricultural practices and responsible stewardship of the land have remained on family farms, moved back after a period of roaming, or settled into country life for the first time. Elizabeth and I feel privileged to count some of these outstanding people as our friends and to tell some of their stories.

ACKNOWLEDGMENTS

After a book is published, it's easy to take all the credit when someone gives a compliment. We'd love to say, "Thanks, we did it all by ourselves!" But books don't get completed that way. Just as farming in upstate New York requires perseverance, determination, and the combined efforts of many individuals, so does the production of a book. It's a team effort.

We do not travel home much in the winter months. As a consequence, we had no suitable images to use for the paintings in the winter section. Fortunately, three photographers—Jon McClellan, Bob Olbrycht, and Karen Christensen—shared their photos.

Elizabeth photographed several scenes for the cover art, none of which seemed quite right. We drove up and down back roads in all kinds of weather, with Elizabeth frequently asking to stop on particularly steep grades for a shot out the car window. We scouted the region and photographed many farms and roadside markets in spring, summer, and fall, all to no avail. Finally, we appealed to our friends Meg and Rob Southerland, and they e-mailed us an absolutely beautiful photo of their farm, Gardenworks, taken by Pete Hanks.

Many people ask, "How do you publish a book?" The answer is, "It requires quite a few steps." After Elizabeth has completed several paintings, she lugs them to a photographic studio, where a professional photographer takes a high-resolution photo of each painting. Each and every image must be just right; the color mix, contrast, tone, and density in the photo have to match the colors in the original artwork exactly. Elizabeth is a perfectionist, so I am sure she put the photographer through his paces. Many thanks to F. Paul Galeone Photographers in Timonium, Maryland, for an excellent job! We also gratefully acknowledge our web designers at Tech Team, who handle our web site and a host of other administrative matters.

We also want to thank our friends, family, and dogs, who never once complained when we said, "We can't go out now, or come for dinner, or throw that ball!" Now that the book is done, everybody's happy.

And last but not least, special thanks to the team at BookPros, who helped make this project possible.

BIBLIOGRAPHY

Agricultural Marketing Resource Center. "Bees profile." AgMRC. http://agmrc.org/commodities__products/livestock/bees_profile.cfm.

Betterbee. "Honey and Candles." http://www.betterbee.com/departments.asp?dept=148.

Cornell Sugar Maple Research & Extension Program. "Frequently asked questions for the Maple Producer: Maple Syrup Collecting and Processing." Cornell University. http://maple.dnr.cornell.edu/FAQ.htm#FAQ2.

Current, Richard Nelson. Wisconsin: A History. New York: Norton, 1977. 75–76.

Feiger, Dale. "Farmacology." Johns Hopkins Magazine vol. 61 no. 3 (2009): 25–31.

The Great Idea Finder. "Potato Chip History – Invention of Potato Chips." http://www.ideafinder.com/history/inventions/potatochips.htm.

Histoire en Ligne. "Antoine-Augustin Parmentier." http://www.histoire-en-ligne.com/article.php3?id_article=301.

National Corn Growers Association. "Research & Business Development." NCGA. http://ncga.com/research-business-development.

Prince Edward Island Potato Museum. "The Potato Industry: Alive and Well in O'Leary, PEI." Virtual Museum Canada. http://www.virtualmuseum.ca/pm.php?id=record_detail&fl=o&lg=english&ex=00000121&rd=55571.

Reisman, Nick. "Fearing prosperity Farmers worry development could drive them out." Glen Falls Post-Star. May 31, 2009. http://www.poststar.com/news/local/article_ae4fb51e-e3ce-5245-b942-80585666db4a.html.

Shaw, Thomas. Soiling Crops and the Silo. New York: Orange Judd Company, 1903.

UPSTATE NEW YORK

•Towns That We Love•

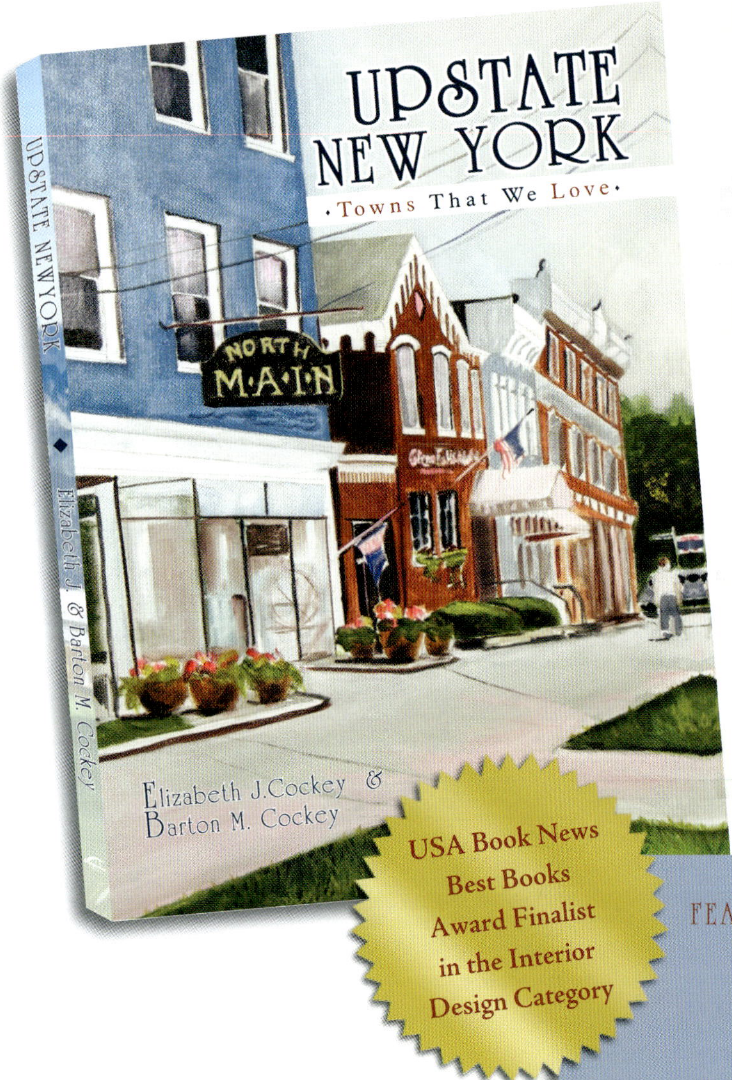

UPSTATE NEW YORK is an illustrated history of the countryside in the Upper Hudson River Valley, including six towns that played an important role in the American Revolution. It is a "travel" book that discusses the life and times of the people who live there now and who lived there in days gone by.

"Elizabeth and Barton Cockey reveal fascinating and alluring aspects of these wonderful places through art and text. Here are quintessential images of our homes and farms, our Main Streets, our treasured landscapes, our storied past, and much, much more. This volume welcomes everyone to our towns, our places—rich in flavor, rich in history, rich in lore. This book will show and tell you why we who live here, or once lived here, will always call it home."

—Tim Tefft, Greenwich, New York,
author, historian, editor, and publisher

FEATURING THESE HISTORIC TOWNS:

Greenwich • Easton
Schuylerville • Salem
Saratoga Springs • Cambridge

and other famous places in the Upper Hudson River Valley

How to order this book:

Bridgeway Books are distributed to the trade by:

Ingram Book Group Phone: (800) 937-0152 Website: www.ingrambook.com
Baker & Taylor Phone: (800) 775-1800 E-mail: btinfo@btol.com Website: www.btol.com

For more information, please visit www.ThePaintedWord.info.